Changing Calculus

A Report on Evaluation Efforts and National Impact from 1988 to 1998

©2001 by
The Mathematical Association of America (Incorporated)
Library of Congress Catalog Card Number 2001091661

ISBN 0-88385-167-9

Printed in the United States of America

Current printing (last digit):
10 9 8 7 6 5 4 3 2 1

Changing Calculus

A Report on Evaluation Efforts and National Impact from 1988 to 1998

Susan L. Ganter
Clemson University

Published and Distributed by
The Mathematical Association of America

The MAA Notes Series, started in 1982, addresses a broad range of topics and themes of interest to all who are involved with undergraduate mathematics. The volumes in this series are readable, informative, and useful, and help the mathematical community keep up with developments of importance to mathematics.

MAA Notes

11. Keys to Improved Instruction by Teaching Assistants and Part-Time Instructors, *Committee on Teaching Assistants and Part-Time Instructors, Bettye Anne Case,* Editor.

13. Reshaping College Mathematics, *Committee on the Undergraduate Program in Mathematics, Lynn A. Steen,* Editor.

14. Mathematical Writing, by *Donald E. Knuth, Tracy Larrabee, and Paul M. Roberts.*

16. Using Writing to Teach Mathematics, *Andrew Sterrett,* Editor.

17. Priming the Calculus Pump: Innovations and Resources, *Committee on Calculus Reform and the First Two Years,* a subcommittee of the Committee on the Undergraduate Program in Mathematics, *Thomas W. Tucker,* Editor.

18. Models for Undergraduate Research in Mathematics, *Lester Senechal,* Editor.

19. Visualization in Teaching and Learning Mathematics, *Committee on Computers in Mathematics Education, Steve Cunningham and Walter S. Zimmermann,* Editors.

20. The Laboratory Approach to Teaching Calculus, *L. Carl Leinbach et al.,* Editors.

21. Perspectives on Contemporary Statistics, *David C. Hoaglin and David S. Moore,* Editors.

22. Heeding the Call for Change: Suggestions for Curricular Action, *Lynn A. Steen,* Editor.

24. Symbolic Computation in Undergraduate Mathematics Education, *Zaven A. Karian,* Editor.

25. The Concept of Function: Aspects of Epistemology and Pedagogy, *Guershon Harel and Ed Dubinsky,* Editors.

26. Statistics for the Twenty-First Century, *Florence and Sheldon Gordon,* Editors.

27. Resources for Calculus Collection, Volume 1: Learning by Discovery: A Lab Manual for Calculus, *Anita E. Solow,* Editor.

28. Resources for Calculus Collection, Volume 2: Calculus Problems for a New Century, *Robert Fraga,* Editor.

29. Resources for Calculus Collection, Volume 3: Applications of Calculus, *Philip Straffin,* Editor.

30. Resources for Calculus Collection, Volume 4: Problems for Student Investigation, *Michael B. Jackson and John R. Ramsay,* Editors.

31. Resources for Calculus Collection, Volume 5: Readings for Calculus, *Underwood Dudley,* Editor.

32. Essays in Humanistic Mathematics, *Alvin White,* Editor.

33. Research Issues in Undergraduate Mathematics Learning: Preliminary Analyses and Results, *James J. Kaput and Ed Dubinsky,* Editors.

34. In Eves' Circles, *Joby Milo Anthony,* Editor.

35. You're the Professor, What Next? Ideas and Resources for Preparing College Teachers, *The Committee on Preparation for College Teaching, Bettye Anne Case,* Editor.

36. Preparing for a New Calculus: Conference Proceedings, *Anita E. Solow,* Editor.

37. A Practical Guide to Cooperative Learning in Collegiate Mathematics, *Nancy L. Hagelgans, Barbara E. Reynolds, SDS, Keith Schwingendorf, Draga Vidakovic, Ed Dubinsky, Mazen Shahin, G. Joseph Wimbish, Jr.*

38. Models That Work: Case Studies in Effective Undergraduate Mathematics Programs, *Alan C. Tucker,* Editor.

39. Calculus: The Dynamics of Change, *CUPM Subcommittee on Calculus Reform and the First Two Years, A. Wayne Roberts,* Editor.

These volumes can be ordered from:
MAA Service Center
P.O. Box 91112
Washington, DC 20090-1112
800-331-1MAA   FAX: 301-206-9789

Acknowledgements

The author would like to thank John Luczak, Patty Monzon, Natalie Poon, and Joan Ruskus for their assistance in compiling the information for this report.

The data collection and subsequent analysis for this study were completed while in residence at the National Science Foundation (NSF) as a Senior Research Fellow of the American Educational Research Association (AERA) and with support from both AERA and NSF. Results reported here are the work of the author and do not reflect the opinion of either AERA or NSF.

Contents

Illustrations

Introduction and Background

More than 500 mathematics departments at postsecondary institutions nationwide are currently implementing some level of calculus reform. These "reformed" courses are affecting an estimated 300,000 calculus students each year, approximately 32% of the students enrolled in calculus (Tucker & Leitzel, 1995).[1] With such tremendous growth of an effort begun only a decade ago, it is likely that calculus reform will continue to affect an even greater percentage of students (Foley & Ruch, 1995; George, 2000; Lightbourne, 2000; Roberts, 1996; Schoenfeld, 1996; Smith, 1996, 2000).

Many institutions nationwide have implemented programs to improve learning in science, mathematics, engineering, and technology (SMET), including some that are working to eliminate the traditional boundaries between these disciplines to produce a truly integrated teaching approach. These programs represent a change in the fundamental philosophy that has long guided the structure of undergraduate education, and it is believed by many that such change is necessary for students who will live and work in an increasingly technical society. However, there are a limited number of studies that document the impact of these efforts on student learning, faculty and student attitudes, and the general environment at undergraduate institutions. It is critical that such studies be conducted so that the academic community can understand the impact of this change in philosophy on learning within and across disciplines, throughout a student's experience at the undergraduate level, and beyond.

Such a study has been conducted on calculus reform as a part of a larger effort by the National Science Foundation (NSF) to evaluate the impact of reform in SMET education at the undergraduate level. This study was designed to investigate what was learned in the ten years from 1988 to 1998 about the effect of calculus reform on (1) student achievement and attitudes, (2) faculty and the mathematics community, and (3) the general educational environment.

A. The history and vision of calculus reform

The need to re-think the undergraduate mathematics curriculum has been widely discussed for more than a century (e.g., Durell, 1894), with a handful of individuals addressing this need in the 1950s and 1960s with new curricula and teaching methods (e.g., Baum, 1958; CUPM, 1969; Davidson, 1970; Davis, 1966; Holden, 1967; Levi, 1963; McKeachie, 1954). In the mid-1980s, undergraduate SMET education began receiving renewed national attention through discussion groups and workshops at which leaders from individual SMET disciplines convened. These meetings resulted in a variety of reports that emphasized the need to support reform in undergraduate SMET education, including the Neal Report in 1986 (NSB, 1986).

In response to this need, NSF revived the Division of Science, Engineering and Mathematics Education, now known as the Division of Undergraduate Education (DUE). A major charge of DUE is to support and

---

[1] In discussions with Tucker and Leitzel, they indicated that they directly contacted a representative sample of non-respondents from their survey, with comparable results.

promote the development of new SMET curricula, pedagogical methods, and educational materials at a wide variety of undergraduate institutions (NSF, 1996). Initiatives resulting from this charge have addressed

- calculus (1988)
- engineering (1989)
- general curriculum development in SMET disciplines (1991)
  (currently the Course, Curriculum, and Laboratory Improvement (CCLI) program)
- bridges to and from calculus (1992)
- teacher preparation (1992)
- chemistry (1994)
- technology (1994)
  (currently the Advanced Technological Education (ATE) program)

and, more recently

- mathematics across the curriculum (1995)
- institution-wide reform (1996), and
- recognition awards for the integration of research and education (1997).[2]

Other agencies have also answered the call to support reform through programs such as the Fund for the Improvement of Post-Secondary Education (FIPSE) at the U.S. Department of Education and individual grants from private organizations such as the Sloan and Exxon Foundations.

Undergraduate institutions must examine the reform efforts of the past decade and capitalize on the methods and curricula that make SMET literacy a part of every graduate's knowledge base. As early as 1991, NSF began receiving pressure from the academic community and Congress to place more emphasis on evaluating the impact of these developments on student learning and the environment in undergraduate institutions (see NSF, 1991). Although this pressure has resulted in a heightened awareness of the need for evaluation and financial support for a few such studies, the area of evaluation research in undergraduate reform is still in its infant stages, with much of the work done by graduate students in the form of unpublished doctoral dissertations and master's theses. Only through additional evaluation research can the SMET community truly understand the impact of their efforts and, consequently, make informed decisions about the future of undergraduate SMET education (AMATYC, 1995; Asiala, et al., 1996; Atkins, 1994; Bookman, 2000; Dossey, 1998; Dudley, 1997; Frid, 1994; Hurley, Koehn, & Ganter, 1999; NSF, 1996; Stevens, Lawrenz, & Sharp, 1993; Tucker & Leitzel, 1995).

B. Evaluation and SMET reform

Evaluation is defined to be the "process of determining the merit, worth and value of things." (Scriven, 1991, p. 1) In a setting such as academia, evaluation becomes an analytical process that enables instructors and administrators to understand the educational system not only within individual disciplines, but across the broad range of multi-disciplinary perspectives within the institution that combine to form a student's individual experiences. Therefore, an effective evaluation study must not only gather and summarize the data needed for decision making, but also interpret and report on the implications of such data based on relevant values and standards (Scriven, 1991).

The evaluation of programs in SMET education was first initiated as a result of the Sputnik era. The development of new SMET courses became prominent, and with these courses came the need to evaluate their impact on the training of future scientists. Unfortunately, the evaluations conducted were not well received by the scientific community, as was articulated by Cronbach in 1963. Cronbach stated that in order

---

[2] Dates in parentheses indicate the fiscal year in which these DUE programs first made awards.

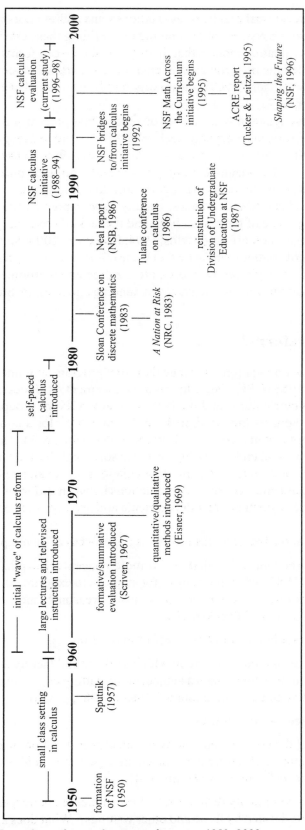

Figure 1. *Abbreviated timeline for undergraduate reform in mathematics, 1950–2000.*

to be effective in shaping educational practices, evaluations must address not only the products coming out of the reform efforts, but also the process of developing and refining these curricula and products. This led to the development of new evaluation models that distinguished between formative and summative evaluation (Scriven, 1967), promoted the importance of evaluating instructional objectives through qualitative, as well as quantitative, measures (Eisner, 1969), and incorporated the notion of evaluator as judge and enforcer of values and beliefs (Eisner, 1975).

Recent reform efforts have resulted in a number of reports that present programmatic information and indicators of success in the efforts to incorporate technology and sound pedagogical methods in undergraduate SMET courses (e.g., NSF, 1993, 1995; Roberts, 1996; Solow, 1996; Tucker & Leitzel, 1995). Reform has received mixed reviews, with students seemingly faring better on some measures, while lagging behind students in traditional courses on others. These reports still do not fully answer the many questions about student learning in undergraduate SMET courses. Faculty have expressed the need for such information, while also realizing the magnitude of the task at hand and the lack of concrete guidelines on how to implement meaningful evaluations of their efforts (Atkins, 1994; Bookman, 2000; Slavin, 1999; Tucker & Leitzel, 1995; West, 1994). The lack of studies to indicate that these efforts are having a positive impact on students, and the increase in workload brought on by reform, are creating an environment of uncertainty that could result in the withdrawal of support for such activities by funding agencies, institutions, faculty and students.

C. What is calculus reform?

In 1987, the National Science Foundation published their first program announcement for calculus, with the initial awards beginning in 1988 (NSF, 1987). In this announcement, NSF described the "need for revision and renewal in the calculus curriculum" and the fact that many relevant groups had begun to pressure the mathematics community to address this need, making the time "ripe for a concerted effort to address the issues of curriculum development in a coherent fashion." (NSF, 1987, p. 1) The focus of this effort was to be on "raising students' conceptual understanding, problem solving skills, analytical and transference skills, while implementing new methods to reduce tedious calculations," with an emphasis on "coordination with other disciplines in science and engineering and interactions between different sectors of the mathematical sciences community in the development of model curricula and prototypical instructional materials." (NSF, 1987, p.1)

Other publications have described calculus reform as the process of developing a course that is

> "leaner, contain[s] fewer topics, but that...ha[s] more conceptual depth, numerically and geometrically," (Douglas, 1987) as "opportunities for students to experiment, to consider larger and more open-ended questions, to work in groups, to justify reasonableness of answers, to encounter realistic problems, to write." (Tucker, 1990)

The movement itself was described in 1987 as an effort to develop

> "a vision of the future of calculus, a future in which students and faculty are together involved in learning, in which calculus is once again a subject at the cutting edge—challenging, stimulating, and immensely attractive to inquisitive minds." (Steen, 1987, p. xiii)

Ten years later, it was still seen as an effort to

> "dramatically chang[e] the content and quality of calculus instruction in all sections of calculus at all types of institutions in all portions of the country [with the goal of] improv[ing] calculus instruction for all students." (Haver, 1998, pp. 3–4)

How has the mathematics community responded to these immense challenges? Do the efforts in calculus reform really reflect the goals defined by these and other writings? What does a "typical" project look like? Specifically, to what did those institutions that received NSF funding choose to devote their efforts?

C.1 *Themes of reform*

An analysis of the files for the NSF-funded projects reveals that the most prominent themes were
- the development of original curricular materials;
- the use of alternative learning environments, such as computer laboratory experiences, discovery learning, and technical writing (each reported by approximately 55% of the projects); and,
- an emphasis on different student skills, such as computer use, whether in a lab setting or in the classroom (appearing in 90% of the funded projects), the use of applications (73%), and a focus on conceptual understanding (64%).

Table 1 outlines some key characteristics of NSF-funded calculus projects (see Chapter 3 for an in-depth discussion about the nature of NSF-funded projects).

Table 1. *Characteristics of NSF-funded calculus projects.*

Characteristic	Frequency ($n = 127$)	Percentage (%)
Carnegie classification		
Research/doctoral	73	57.4
Comprehensive	19	15.0
Liberal arts	19	15.0
Two-year	11	8.7
Other	5	3.9
Special institution type*		
Engineering	8	6.3
Women only	3	2.4
Historically black colleges and universities	3	2.4
Project type**		
Original curriculum development	74	58.2
Implementation	30	23.6
Workshops	15	11.8
Conferences	5	3.9
Newsletters, videos, other dissemination	3	2.5
Project sub-type*		
Software development	32	25.1
Links to other institution(s)	26	20.4
Planning grant	20	15.7
Target audience**		
College students	123	96.8
High school students	59	46.4
Minority students	44	34.6
Community college students	34	26.7
Female students	26	20.4
Engineering students	14	11.0
Preservice teachers	12	9.4
Life science students	7	5.5
Business students	6	4.7
Disabled students	5	3.9
Non-math majors	4	3.1

*A project will not necessarily contribute to any category.
**A project may contribute to more than one category.

C.2 *Student audiences and special institutions*

Almost all funded institutions (97%) indicated that their reform efforts were targeted at college students. However, a significant number of projects indicated that they were also focusing on the needs of high school students (46%). Within the college population, minorities were the subgroup that received the most attention (35%), although several other groups were targeted including community college students (27%), engineering majors (11%), and preservice teachers (9%). Surprisingly, women were a primary focus for only 20% of the efforts, many of which were at women's colleges (see Table 1 and Chapter 3, Section D).

C.3 *Evaluation and dissemination*

Almost half (48%) of the funded projects reported the use of some evaluation strategy as part of their work. The most frequently cited methods for collecting evaluation data were student surveys (33%), achievement exams/GPAs (27%), retention/completion rates (21%), and experimental vs. control studies (21%). Methods for disseminating materials and ideas that were developed as part of the projects were quite conventional, with the most frequent being conference presentations (58%), journal articles (43%), and invited colloquia (36%) (see Table 8 and Chapter 3, Section E).

D. The need for this study

Research on the teaching and learning of calculus as influenced by the reform movement is very new, with only a few studies that investigate student achievement, the general state of calculus instruction, or the influences of non-traditional curricula and teaching strategies in calculus on student learning (Asiala, et al., 1996; Bookman & Friedman, 1994; Frid, 1994; Ganter, 1999; Schoenfeld, 1996; Tucker & Leitzel, 1995). Many studies that are available have been conducted only very recently as graduate research, with the results published as theses and dissertations and therefore not widely disseminated. This is similar to the trend that was seen thirty years earlier, when studies to evaluate the impact of reform were not supported to the extent necessary to adequately inform progress. Consequently, there is a great need for information on various instructional formats and their subsequent effects on learning.

In addition, the mathematical community needs to develop an understanding of the perspectives, methodologies, and findings of research on calculus learning, as it is important to discover the "right mix" of content and pedagogy (Crockett & Kiele, 1993; Ferrini-Mundy & Lauten, 1994; Tucker & Leitzel, 1995). Concrete information about what and how students learn, as well as decisions about program improvements, cannot be reasonably made with only anecdotes and success stories. The value of changes made to a course with such impact as calculus is very difficult to quantify; such decisions cannot be made without an enormous amount of data collected over a long period of time that include information on students before, during, and after the course. Ultimately, the true goal of any calculus course is to ensure that students completing the course will be able to adequately analyze and solve problems involving calculus, whether those problems be in physics, engineering, other mathematics courses, or the workplace (Davis, 1993, 1994, 2000; Gordon, 2000).

It is important to begin this process not only by conducting evaluation studies, but also by reviewing the relevant literature and assessing what is already known about the impact of reform efforts. Although there has been considerable work in the area of student learning in calculus (e.g., Davidson (1990), Heid (1988), and Tall (1990)), most of these investigations have looked at how students learn. It is imperative to understand not only how students learn, but also the actual impact of different environments on their ability to learn. This project, sponsored by the American Educational Research Association (AERA) and the National Science Foundation (NSF), focuses on the collection, analysis, and dissemination of such information as impacted by the calculus reform movement.

E. Organization of the report

The remainder of this report begins with a discussion of the study design, including methodology, organization of the data collection, and the model used to analyze the information. Results are then presented from studies in three areas: student achievement, student attitudes, and faculty reactions. Following this are the discussion, conclusions and implications for further research. An extensive bibliography of references relevant to the study of calculus has been assembled and organized by topic to assist in the initiation of these subsequent studies.

Design of the Study

A. Methodology for the study

The calculus reform efforts that NSF encouraged through awards from 1988 to 1994 set the direction for much of the undergraduate reform that has followed. A synthesis of evaluations implemented as part of these projects and others in calculus between 1988 and 1998 has been conducted. Specifically, this study was designed to obtain information about the impact of calculus reform on student achievement and attitudes, faculty and the mathematics community, and the general learning environment using the following sources:

- Files containing all information that has been submitted to NSF for each NSF-funded calculus project were searched for any proposed evaluation of the project and corresponding findings, as well as dissemination information. Results of the search were compiled in a qualitative database.
- A search of the literature was conducted, including published journal articles, conference proceedings, dissertations, and other scholarly works, to identify evaluations of calculus reform.
- Responses to a letter sent to approximately 600 individuals who have contributed to the reform efforts were reviewed. The letter requested assistance in the compilation of existing evaluation studies in calculus reform.

B. Development of the database

Prior to reviewing NSF files for the 127 projects funded in the calculus program between 1988 and 1994, a framework was developed in consultation with NSF staff, as well as scientists from SRI International who were assisting with the compilation of the data (see Appendix A). The framework was used to guide the search of NSF files, as well as in the development of the database for compiling the results of the search. Precise definitions of key words to be used in determining the existence of various evaluation, dissemination, and reform activities for each project were also developed and used to guide the search and data entry. This level of precision meant that several individuals could be utilized to search the enormous amount of information in the files without loss of consistency in the data entered.

The information was summarized in a computerized database that contains individual records for each of the 127 projects, creating the ability to analyze outcomes across projects. The database was developed using *Access,* a qualitative data analysis program. A qualitative database was chosen to maximize the amount and types of information that could be organized from the files. This database allowed not only for statistical tabulations, but also for the synthesis of outcome summaries for each of the projects. For example, findings from evaluation studies for each project were summarized into brief paragraphs. These paragraphs for all projects using, for example, cooperative learning could then be condensed onto one page by the computer, making it possible to synthesize large amounts of outcome information for selected reform methodologies. Similarly, outcomes could be synthesized by any other input variables developed in the framework.

Table 2. *Input and outcome variables used in data analyses, with frequency of NSF-funded projects measuring each outcome variable.*

Input variables	Outcome variables	Frequency (%)*
Carnegie classification	Student attitudes	52 (40.9)
Special institution type	Student achievement	34 (26.8)
Project type	Student retention	26 (20.5)
Target audience	Student majors/degrees	
National objectives of reform *		1 (0.8)
Pedagogical strategies***	Faculty reactions	28 (22.1)
	Cost effectiveness	1 (0.8)

*$n = 127$
**See Chapter 3, Section B for clarification.
***See Chapter 3, Section C for clarification.

Therefore, analyses were conducted on several outcome variables as a function of numerous input variables, including national objectives of reform implemented, target audiences, and Carnegie classification (see Table 2). Results were then aggregated to produce an overall picture of these outcomes, as well as insights about the relative impact of the various input areas on these aggregate outcomes. Note that not all areas explored yielded adequate information to warrant a discussion of the results.

A system also was developed to determine the relative strength of the evaluation studies conducted by the NSF-funded projects. A list of indicators of evaluation quality was devised internally and used to rate the evaluation methodologies on the basis of the number of these indicators implemented as part of the study (see Appendix B). The ratings could then be used to

- develop a sense of the quality of evaluations conducted,
- determine the validity of the overall results by comparing them with results from the "top" evaluations, and
- inform future project evaluations.

This process is discussed in more detail in section D.2 of this chapter.

C. The literature search and letter of request to the community

An extensive search of the relevant literature was conducted. The information was compiled and synthesized to determine the current status of calculus reform, as well as efforts to evaluate these reform efforts. Results from the literature review have been used to develop numerous sections of this report, including the background material in each of the outcome areas, as well as for reform in general. Evaluations that were part of an NSF-funded project were combined with the other information in the project's NSF file, while studies from the search that were not directly attributed to an NSF-funded project were compiled separately and are reported as part of the aggregate outcomes. The literature review was also used to compile a list of individuals who had participated in the reform efforts or who played a significant role in the current or past status of calculus.

Because the evaluation of calculus reform is still in its infant stages, much of the documentation on this and other areas of calculus reform has not been published in conventional ways. In order to fully study the impact of reform efforts, it was necessary to review not only published documents attainable through the literature search, but also the wealth of information available in other forms. Therefore, a letter requesting information on calculus reform and any evaluations pertaining to the impact of those efforts was developed (see Appendix C). The letter was reviewed internally at NSF and then sent via email to a select group of mathematicians and mathematics educators for feedback and comments prior to the mailing. This "heads up" to leaders in calculus reform and the general mathematics community helped to establish discussions about this evaluation effort, ultimately leading to stronger support for the project. The letter was mailed in July 1996 to more than 600 individuals from the mailing list developed as part of the literature search. As previously discussed, this mailing list included not only principal investigators from NSF calculus projects, but also individuals from non-funded efforts and many others who have been involved in other aspects (pro and con) of calculus reform.

D. Scope of information

D.1 *Quantity and type*

NSF files for all 127 projects awarded to 110 institutions as part of the calculus program in the Division of Undergraduate Education (1988–94) were obtained. Each of these files was extensively reviewed as described above and condensed into folders containing only the information relevant to this study. This information was then entered into the database using the methodology described. Although each file was carefully examined and every effort was made to obtain complete information on all projects, aggregate outcomes are conservative estimates, as many of the NSF files were incomplete.

The literature search and letter to the community yielded 163 documents on calculus reform in addition to those directly related to the NSF-funded projects. These documents include published journal articles, reports and collections of articles, papers from conference proceedings, unpublished manuscripts, newsletters, doctoral dissertations, masters theses, internal institutional reports, and letters from individuals. This collection of materials spans current and past reform efforts, including

* corresponding evaluations
* expository writings on calculus reform
* outcomes from conferences, workshops, and meetings devoted to calculus reform; and,
* anecdotal discussions from individual faculty about the impact of calculus reform in their departments and institutions.

Although it would be inaccurate to claim that the search was exhaustive, the aggregate materials collected through the review of NSF files, literature review, and letter to the mathematics community are quite extensive and certainly provide the background necessary to motivate informed discussions about the impact of calculus reform.

Table 3. *Sources of evaluation data.*

Source	Frequency (%)*		
	NSF-funded projects	**Non NSF-funded projects**	**Total**
Published article	43 (13.6)	72 (22.8)	**115 (36.4)**
Published report/collections of articles	0 (0.0)	16 (5.1)	**16 (5.1)**
Conference proceedings	29 (9.2)	20 (6.3)	**49 (15.5)**
Unpublished manuscript	23 (7.3)	8 (2.5)	**31 (9.8)**
Newsletter	2 (0.6)	0 (0.0)	**2 (0.6)**
Doctoral dissertation	11 (3.5)	21 (6.6)	**32 (10.1)**
Masters thesis	2 (0.6)	3 (0.9)	**5 (1.6)**
Internal institutional report or report to NSF	16 (5.1)	2 (0.6)	**18 (5.7)**
Individual letters	27 (8.5)	21 (6.6)	**48 (15.2)**
Total	**153 (48.4)**	**163 (51.6)**	**316 (100.0)**

*n = 316 for all percentage calculations.

D.2 *Quality of data for this report*

In the first several years of the NSF calculus initiative, the focus was on the development and dissemination of innovative curricular materials. As a result, a wide variety of materials ranging from textbooks to computer laboratory guides to collections of long-term projects were made available to the mathematics community very quickly. This strong emphasis on curriculum development necessarily implied that the evaluation of the impact of these materials in the classroom and on the culture within departments was virtually non-existent. By the early 1990s, it became clear that documentation of this impact was an important part of the reform process. Many individuals in the mathematics community who were unconvinced that the reform methods were a positive addition to the curriculum began demanding evidence to support the claims of those involved in reform. Increasing pressure on NSF to be accountable for the expenditure of funds also heightened the awareness of the need for evaluation. As a result, projects funded in the later years of the calculus initiative, as well as non-funded efforts during this later time period, were more likely to include an evaluation plan as part of the implementation process.

Although the decision to place a greater emphasis on assessing the impact of reform efforts through evaluation came a bit late, it was nonetheless a good one. In spite of the good intentions of the reform community, the decision was made without a consistent plan for how these evaluations would be conducted. Most of the project leaders had no experience in designing evaluation studies and, not surprisingly, investigated the impact of their work in ways that resulted in unreliable information; e.g., comparing the course grades of traditional and reform instructors. As these project leaders became more experienced, many began to seek the advice and assistance of evaluation researchers. However, even this important step was often taken in a superficial manner, with an evaluation "plan" simply added to a curriculum development project as an afterthought rather than as an integral part of the implementation process. This again resulted in the collection of data that was unreliable or inconsistent with the goals of the project, if in fact specific goals had been developed at all.

This learning process on the part of the mathematics community, as well as NSF, has ultimately resulted in a greater understanding and appreciation of the development and implementation of reliable evaluation studies, including consultation with evaluation experts, as part of the reform process. This understanding has come only after many years of experimentation and the loss of an enormous amount of information that is critical to fully analyzing the impact of calculus reform. Therefore, it was necessary to develop a method for assessing the perceived quality of the evaluation studies conducted as part of the NSF-funded projects.

A list of indicators of evaluation quality was developed in consultation with evaluators from SRI International (see Appendix B). Each of these indicators was carefully defined, including methods for determining which indicators had been implemented in a given evaluation study. Each study was then awarded a "rating" based on the number of indicators implemented. Fortunately, there are projects (although limited in number) for which reliable evaluation information has been collected, some over extended periods of time. Only evaluations rating at least 9 out of 14 ($n = 9$, or approximately 7% of the projects) were considered strong evaluations (see Table 4).

Table 4. *Distribution of quality ratings for NSF-funded project evaluations used to validate results of aggregate data analyses.*

Primary institution	Rating (out of 14)
Duke University	12
Purdue University	12
New Mexico State University	11
Oregon State University	9
University of Illinois at Urbana-Champaign	9
University of Iowa	9
University of Michigan, Ann Arbor	9
University of Tennessee at Chattanooga	9
Worcester Polytechnic Institute	9

These studies can be used to corroborate or refute the claims of other evaluation results that have been obtained using less rigorous methodologies; that is, to "validate" the results of the entire pool. Results from each outcome area that were observed for the entire pool of information, including both projects with and without NSF support, were compared with the corresponding results from these stronger evaluation projects. Only results that were supported by this corroboration will be discussed in this report. Even with this attempt to validate the data, it should be noted that all results to be reported are suspect, since no overall evaluation design was used to collect even the stronger data in a manner that is consistent across projects.

E. Reporting of the data

Because the nature of the data collected in this study is different from that typically collected in an evaluation study, the method used to report the data is also unconventional. In fact, the study is actually a synthesis of data already collected as part of previously conducted studies and informal inquiries. Therefore, this report is not intended to be a complete investigation of the impact of calculus reform, but rather a compilation of what was learned by the community as a whole between 1988 and 1998.

Four types of information are reported:

- general NSF results
- student achievement
- student attitudes, and
- faculty reactions.

With the exception of the general NSF results, each of these sections reports information from both NSF-funded projects and results from work at other institutions. Except for a reference to the number of each type of study, the results from these two groups are not distinguished. This method of reporting was done in an effort to synthesize the overall impact of calculus reform, not simply the work done at NSF-funded institutions.

Of course, there have been many other components of the educational environment that have been affected by calculus reform, such as student achievement in subsequent courses, departmental budgets, and enrollments in various major programs. The areas that are addressed by this report are the only ones for which sufficient information was available to make any reasonable conclusions. In fact, one of the goals for this project was to determine areas in need of further investigation (see Chapter 7, Section B for this discussion).

Statistical Information on NSF-funded Projects

Although calculus reform efforts have not been limited exclusively to projects supported by NSF, many have been directly influenced by this funding due to the heavy emphasis on dissemination of the ideas developed in these projects. Therefore, it is helpful to examine the nature of the NSF-funded projects, as they are largely representative of the entire calculus reform movement. The information presented here was compiled from NSF files, as well as other documents submitted by leaders of the projects. For the purpose of this study, a project was counted in a certain category if the existence of that idea was discussed anywhere in the project's documentation, including the grant proposal, progress reports, final reports, or other writings about the project. The actual degree of implementation of these ideas as part of the project was not factored into these results.

A. Types of projects

The NSF calculus initiative primarily supported the development of original curricular materials, as is evidenced by the number of projects of this type (58.2%). The other 41.8% of the projects consisted of implementation projects (23.6%) and the organization of workshops and conferences (15.7%), with the remaining 2.5% devoted to distribution of newsletters, videos, and other such materials (see Table 1). Implementation projects, that is, funded projects that focused on the implementation of materials developed at other institutions, most frequently utilized the calculus curriculum developed by the Harvard consortium ($n = 13$), with several others implementing materials from Duke University ($n = 8$), St. Olaf College ($n = 4$), University of Iowa ($n = 3$), Oregon State University ($n = 3$), Clemson University ($n = 2$), New Mexico State University ($n = 2$), Purdue University ($n = 2$), and University of Illinois ($n = 2$).[3] Although not part of the NSF initiative, the cooperative learning model developed at University of California, Berkeley (see Treisman, 1985) was also utilized by three of the implementation sites. Interestingly, the number of implementation projects using the various curricular materials closely reflects the relative distribution of these materials nationally (see Table 5).

B. National objectives of reform

A list of eighteen national objectives of calculus reform was formulated based on a number of publications that discuss the development of the reform movement (Roberts, 1996; Schoenfeld, 1996; Steen, 1987; Tucker, 1990; Tucker & Leitzel, 1995). These objectives focus on issues related to content and pedagogy that have been defined as the basis, in varying combinations, for all reform efforts (see Appendix D and Table 6). It is

---

[3] Note that several implementation projects utilized combinations of more than one curricular project. Therefore, the total number in this list is greater than the total number of implementation projects (see Table 5).

Table 5. *Calculus reform materials used by other NSF-funded campuses (implementation grants), with name(s) of primary project director(s).*

Location(s) of original curriculum development	Frequency (*n* = 30)	Percentage (%)
Harvard Calculus Consortium (Hughes-Hallett, Gleason, et al.)	6	20.0
Duke University (Moore & Smith)	4	13.3
St. Olaf College (Ostebee & Zorn)	3	10.0
University of Iowa (Stroyan)	3	10.0
Clemson University (Kenelly)	2	6.7
University of California, Berkeley (Treisman)	2	6.7
Purdue University (Dubinsky)	1	3.3
University of Illinois at Urbana-Champaign (Uhl)	1	3.3
Combinations used in NSF-funded implementation grants		
Harvard and Duke	2	6.7
Harvard and Ithaca College (Hilbert, et al.)	1	3.3
Harvard and University of California, Berkeley	1	3.3
Harvard, Duke, and Oregon State University (Dick & Patton)	1	3.3
Iowa State (Johnston), New Mexico State (Cohen, et al.), and Cornell (Livesay)	1	3.3
Harvard, Oregon State, and several other projects	2	6.7

important to understand which issues have in reality been the driving force, as it will help us better understand the nature of calculus reform as defined by the mathematics community.

In reviewing the information for the entire population, several objectives clearly stood out from the rest as the major thrust of the reform efforts. As seen in Table 6, computer use is the most frequently cited component of reform to be planned and/or implemented, appearing in 89.7% of the funded projects. At a distant second with 73.2% of the projects is the use of applications, followed by conceptual understanding (63.7%), laboratory experiences (55.9%), and a tie for fifth by discovery learning and technical writing (55.1%). If we consider the fact that laboratory experience implies computer use in most cases, we can look further into the list to cooperative learning and multiple representations of one concept, both coming in at 53.5% of the funded projects. In any case, it is very clear that this list of the most frequently discussed objectives (which includes all objectives that were mentioned in more than 50% of the projects) is very "top heavy," with the top three (computer use, applications, and conceptual understanding) being utilized at many more institutions than any of the others. In fact, the difference is so great that this observation can be stated with confidence in spite of the weaknesses inherent in the data.

Assuming then that computer use, applications, and conceptual understanding are considered the cornerstones of calculus reform by the mathematics community at large, let us examine these data more closely for various institution types. Table 7 displays the list of eighteen objectives with their corresponding percentages by Carnegie classification. It is of interest to note that computer use remains far and away the most frequently cited component of reform when observing the NSF-funded liberal arts, comprehensive, and research colleges and universities, which all have counts greater than 90%. It is at the two-year institutions that computers seem to have had less influence, although computers still score a very high second (81.8%) behind use of applications (90.9%). This lower emphasis on computers, in combination with the fact that graphing calculators had the third highest percentage for the two-year colleges (63.6%) when it didn't even make the top eight for the liberal arts, comprehensive, and research colleges and universities, is likely indicative of the budgetary difficulties faced by these institutions and their students and, perhaps, a lack of technological training for their faculty members.

With the exception of one other interesting case (to be discussed here), these top three objectives remained strongly at the top of the list. The exception occurs in the data for the comprehensive universities, where computer use and applications were still the top two, but discovery learning and technical writing scored above conceptual understanding. Because the percentages here are not dramatically different (78.9% versus 73.6%), it is not believed that this change in ordering is significant. It still is interesting to note that discovery learning and technical writing are considered so important at the comprehensive universities, while they had relatively low rankings at all other institution types.

C. Pedagogical strategies

Several of the national objectives of calculus reform specifically focus on pedagogical strategies. These pedagogical strategies were established in a number of early reform publications as supporting, or being "in the spirit of," the reform efforts (e.g., Steen, 1987; Tucker, 1990). Included in the list of pedagogical strategies are

- alternative assessment methods
- cooperative learning
- discovery learning
- extended-time projects
- laboratory experience
- oral student presentations, and
- technical writing (see Appendix D).

Laboratory experience was the most frequently utilized strategy (55.9%), followed closely by discovery learning, (55.1%), technical writing (55.1%), and cooperative learning (53.5%) (see Table 6). It is somewhat surpris-

Table 6. *Distribution of national objectives of calculus reform.*

Objective	Frequency ($n = 127$)*	Percentage(%)
Computer use	114	89.7
Applications	93	73.2
Conceptual understanding	81	63.7
Laboratory experience	71	55.9
Discovery learning	70	55.1
Technical writing	70	55.1
Cooperative learning	68	53.5
Multiple representations	68	53.5
Problem-solving skills	63	49.6
Graphing calculators	61	48.0
Real-world modeling	61	48.0
Extended-time projects	41	32.2
Approximation	37	29.1
Differential equations	35	27.5
Open-ended problems	31	24.4
Alternative assessment	22	17.3
Mathematical reading	17	13.3
Oral student presentations	8	6.3

*A project may contribute to more than one objective.

Table 7. *Distribution of national objectives of calculus reform by Carnegie classification.*

Objective	Research/doctoral (n = 73)	Comprehensive (n = 19)	Liberal arts (n = 19)	Two-year (n = 11)	Other (n = 5)
	Frequency (%)**				
Computer use	66 (90.4)	18 (94.7)	18 (94.7)	9 (81.8)	3 (60.0)
Applications	51 (69.8)	16 (84.2)	13 (68.4)	10 (90.9)	3 (60.0)
Conceptual understanding	48 (65.7)	14 (73.6)	14 (73.6)	4 (36.3)	1 (20.0)
Laboratory experience	41 (56.1)	11 (57.8)	13 (68.4)	4 (36.3)	2 (40.0)
Discovery learning	41 (56.1)	15 (78.9)	10 (52.6)	3 (27.2)	1 (20.0)
Technical writing	40 (54.7)	15 (78.9)	8 (42.1)	5 (45.4)	2 (40.0)
Cooperative learning	39 (53.4)	13 (68.4)	9 (47.3)	5 (45.4)	2 (40.0)
Multiple representations	39 (53.4)	10 (52.6)	11 (57.8)	6 (54.5)	2 (40.0)
Problem-solving skills	38 (52.0)	11 (57.8)	8 (42.1)	5 (45.4)	1 (20.0)
Graphing calculators	35 (47.9)	10 (52.6)	8 (42.1)	7 (63.6)	1 (20.0)
Real-world modeling	37 (50.6)	9 (47.3)	10 (52.6)	2 (18.1)	3 (60.0)
Extended-time projects	24 (32.8)	7 (36.8)	7 (36.8)	2 (18.1)	1 (20.0)
Approximation	24 (32.8)	2 (10.5)	7 (36.8)	4 (36.3)	0 (0.0)
Differential equations	24 (32.8)	2 (10.5)	4 (21.0)	3 (27.2)	2 (40.0)
Open-ended problems	17 (23.2)	6 (31.5)	7 (36.8)	1 (9.1)	0 (0.0)
Alternative assessment	7 (9.6)	5 (26.3)	8 (42.1)	1 (9.1)	1 (20.0)
Mathematical reading	10 (13.6)	1 (5.3)	4 (21.0)	1 (9.1)	1 (20.0)
Oral student presentations	4 (5.5)	3 (15.7)	1 (5.3)	0 (0.0)	0 (0.0)

*Projects in this category were conducted at non-academic organizations (e.g., professional societies).
**A project may contribute to more than one objective, but only to one Carnegie class.

ing that the other pedagogical strategies, which emphasize alternative methods for assessing student learning, were not considered a high priority. There is evidence that the need for alternative assessment measures to accompany the new methods of presentation that were part of reform was realized by the community only after the reform was well underway. This reflects the "just in time learning" theme that seems to be a part of the entire reform effort; that is, materials and alternative methods of delivery were implemented without considering the impact on student assessment, operations of departments, or success in future courses.

D. Target audiences

As expected, almost all institutions (96.8%) indicated that their reform efforts were targeted at college students. However, a number of institutions indicated that several other audiences were also singled out as targets in their reform efforts. Specifically, a significant number of projects indicated that they were focusing on the needs of high school students (46.4%) and minorities (34.6%). Surprisingly, women were not a primary focus of the efforts (only 20.4% reported this group as a target audience), while community college students, engineering majors, and pre-service teachers were of much greater importance in the efforts than expected, with 26.7%, 11.0%, and 9.4% of the projects, respectively, reporting efforts to focus on these groups (see Table 1).

When reviewing the data again by institution type, the results are fairly consistent with the results from the total population. A few exceptions are worth noting.

1. *Research and comprehensive universities are more likely to involve high schools in their efforts than the liberal arts and two-year colleges.* In fact, only 36.3% of the two-year colleges with funded projects

Table 8. *Evaluation and dissemination strategies.*

Strategy	Frequency ($n = 127$)	Percentage (%)	Total for all projects*
Evaluation			
Student surveys	42	33.1	
Student achievement exams and GPAs	34	26.8	
Student retention and completion rates	26	20.5	
Experimental vs. control group study	26	20.5	
Student interviews	25	19.7	
Longitudinal study	25	19.7	
External review team	18	14.2	
Student course evaluations	16	12.6	
Cooperation with school of education or other university evaluation office	14	11.0	
Faculty surveys	11	8.7	
Conference/workshop evaluations	11	8.7	
Classroom observations/videotapes	10	7.9	
Faculty interviews	9	7.1	
Faculty journals	9	7.1	
Graduate student evaluations	9	7.1	
Gender tracking	8	6.3	
Minority student tracking	3	2.4	
Student attendance rates	3	2.4	
Student journals/writing	2	1.6	
Graduation information	1	0.8	
Cost effectiveness	1	0.8	
Dissemination**			
Use at other colleges	34	26.8	1,116
Conference presentations	73	57.5	608
Use at secondary schools	19	15.0	565
Invited colloquia	46	36.2	345
Organize workshops/conferences	47	37.0	265
Journal articles	54	42.5	229
Publication of course materials	33	26.0	71
Textbooks	19	15.0	38
Internal institution publications	8	6.3	35
Software development	25	19.7	34
Dissertations/theses	8	6.3	21
Requests from other institutions	16	12.6	20
Internet and website development	17	13.4	17
Newsletter development/distribution	11	8.7	11
Videotape development/distribution	4	3.2	4
Publication of conference proceedings	4	3.2	4

*This category only applies to dissemination, since each institution cannot have multiple entries in any one evaluation category.

**Frequencies represent the number of distinct occurrences, not total number of distributions; for example, the frequency of textbooks is the total number of distinct textbooks published, not the total number actually distributed/sold.

reported the involvement of high schools, while 57.8% of the comprehensive universities and 49.3% of the research institutions reported such involvement.

2. *Minority student involvement is significantly more prevalent at the two year colleges (72.7%) than at any other type of institution.* This is perhaps due to the different composition of the student population at two-year colleges, but is certainly a noteworthy observation.

3. *Concerted efforts to involve women in reform efforts are infrequent.* This result is primarily due to the lack of such a focus at the research institutions, with only 15.0% reporting an effort to target the needs of female students. As might be expected, the primary efforts to focus on women occurred at the women's colleges, of which there were three in the NSF funding pool. Two-year colleges also made a significant effort to target women, with 36.3% reporting such a focus.

Data from "special" institutions (women's colleges, historically black colleges and universities (HBCUs), and engineering institutions) were also analyzed. Of the 127 awards in the calculus initiative, 14 went to 11 of these institutions: eight awards to engineering institutions and three to each of the other two types (note that multiple awards were made to three of these institutions). Because of the small number of these institutions in the pool of awardees, it is difficult to make any generalized conclusions about the emphasis of their efforts. It would appear that these special institutions focused on the unique attributes of their students in an attempt to maximize the impact of their efforts. For example, 100% of the engineering institutions reported an emphasis on computers and applications. The women's colleges and HBCUs highly emphasized cooperative learning, a strategy that has been shown to be particularly effective with women and minorities. In addition, it is of interest to note that 100% of the women's colleges and HBCUs also reported an emphasis on computers and applications.

E. Evaluation and dissemination strategies

Although evaluation of the impact of reform efforts was downplayed in NSF efforts in the first several years of the initiative, 47.2% of the funded projects still reported the use of some evaluation strategy as part of their work. The most frequently utilized methods for collecting evaluation data were

- student surveys (33.1% of projects)
- achievement exams/GPAs (26.8%)
- retention/completion rates (20.5%), and
- experimental vs. control group studies (20.5%) (see Table 8).[4]

Several strategies were used by multiple projects that were not included in the original list of evaluation activities, but were observed in the analysis of NSF files. These less conventional methods included student attendance rates and faculty journals. Several projects also solicited the help of colleagues from the School of Education and/or the University Evaluation Office.

Methods for disseminating materials and ideas that were developed as part of the projects were quite conventional, with the most frequent being

- conference presentations (57.5%)
- journal articles (42.5%), and
- invited colloquia (36.2%) (see Table 8).

More interesting is an examination of the estimated number of dissemination activities in each of the investigated categories between 1988 and 1998. The 127 projects that received NSF funding reported a staggering

---

[4] See Appendix A for the entire list of evaluation activities investigated.

Table 9. *Distribution of national objectives of calculus reform by NSF funding levels per year.*

Objective	Total NSF funding awarded/ Number of funded projects* (Average NSF funding per project)**										
	1988	1989	1990	1991	1992	1993	1994	1995	1996	1997	Total
Computer use	1118/22 (51)	2098/19 (110)	2633/26 (101)	2888/32 (90)	3661/33 (111)	3002/25 (120)	3343/24 (139)	638/3 (213)	572/2 (286)	299/1 (299)	20252/187 (108)
Applications	114/22 (52)	1925/15 (128)	2478/22 (113)	2517/26 (97)	3220/28 (115)	2655/21 (126)	2380/17 (140)	273/1 (273)	273/1 (273)		1686/153 (110)
Conceptual understanding	750/16 (47)	1629/13 (125)	1900/16 (119)	2026/22 (92)	2998/25 (120)	3241/23 (141)	3044/20 (152)	783/3 (261)	678/2 (339)		1704/140 (122)
Laboratory experience	651/10 (65)	1293/14 (92)	1740/19 (92)	2195/26 (84)	2325/23 (101)	1785/18 (99)	1857/16 (116)	378/2 (189)	273/1 (273)		12498/129 (97)
Discovery learning	586/11 (53)	1449/12 (121)	2278/20 (114)	2130/22 (97)	2599/22 (118)	2914/20 (146)	2450/15 (163)	510/2 (255)	405/1 (405)		1532/125 (123)
Technical writing	513/10 (51)	922/8 (115)	1573/18 (87)	1769/21 (84)	2308/24 (96)	1898/18 (105)	1853/14 (132)				1084/113 (96)
Cooperative learning	469/7 (67)	647/7 (92)	1361/14 (97)	1804/22 (82)	2919/27 (108)	2651/22 (121)	2938/20 (147)	378/2 (189)	273/1 (273)		1344/122 (110)
Multiple representations	545/11 (50)	1405/13 (108)	2103/18 (117)	1847/20 (92)	2912/22 (132)	2560/18 (142)	2588/15 (142)	938/3 (313)	977/3 (326)	299/1 (299)	16174/124 (130)
Problem-solving skills	765/15 (51)	1407/13 (108)	1716/19 (90)	1589/19 (84)	2035/18 (113)	1880/14 (134)	1519/12 (127)	510/2 (155)	405/1 (405)		11827/113 (105)
Graphing calculators	584/12 (49)	947/9 (105)	1583/12 (132)	1649/16 (103)	2708/21 (129)	2594/16 (162)	2249/14 (161)	665/2 (332)	704/2 (352)	299/1 (299)	13982/105 (133)
Real-world modeling	760/13 (58)	1365/11 (124)	1992/22 (91)	1726/18 (96)	1850/19 (97)	1560/14 (111)	1893/13 (146)	273/1 (273)	273/1 (273)		11692/112 (104)
Extended-time projects	360/6 (60)	935/8 (117)	1097/12 (91)	1005/12 (84)	1590/15 (106)	1007/10 (101)	1038/11 (94)				7031/74 (95)
Approximation	570/10 (57)	1248/10 (125)	1654/11 (150)	1484/13 (114)	1818/14 (130)	1532/9 (170)	986/5 (197)	405/1 (405)	405/1 (405)		1010/74 (137)
Differential equations	257/4 (64)	1120/5 (239)	1560/10 (156)	1486/13 (114)	1545/12 (129)	1251/11 (114)	1330/8 (166)	273/1 (273)	273/1 (273)		9173/65 (141)
Open-ended problems	328/4 (82)	919/8 (115)	959/9 (107)	1096/12 (91)	837/10 (84)	687/9 (76)	789/8 (99)				5616/60 (94)
Alternative assessment	161/3 (54)	833/5 (167)	1179/6 (196)	931/8 (116)	882/7 (126)	940/6 (157)	386/2 (193)	273/1 (273)	273/1 (273)		5859/39 (150)
Mathematical reading	187/3 (62)	584/4 (146)	709/6 (118)	449/5 (90)	541/6 (90)	150/1 (150)	605/4 (151)				3225/29 (111)
Oral student presentations	176/3 (59)	99/1 (99)	283/4 (71)	80/1 (80)	320/2 (160)	145/1 (145)	100/1 (100)				1204/13 (93)
Total (actual annual NSF figures)	**1290/25 (52)**	**2273/22 (103)**	**2939/30 (98)**	**3040/35 (87)**	**4075/37 (110)**	**3895/25 (134)**	**2848/26 (148)**	**1043/4 (261)**	**977/3 (326)**	**299/1 (299)**	**23680/212 (112)**

*Total NSF funding for each objective in each year represents the total amount awarded (in thousands of dollars) to institutions indicating that this objective was part of their project; that is, these funds were not necessarily devoted only to this objective, but rather to institutions working on this objective (perhaps in conjunction with other objectives). Therefore, in any given year, some funds are recorded in multiple categories and so the bottom totals (in boldface) do not equal the sum of the corresponding column figures. Although NSF does not require individual projects to itemize budgets based on objectives such as the ones listed here, the distribution of dollars to projects working on such objectives clearly serves as an indicator of funding priorities in each year and trends throughout the NSF calculus initiative.

**NSF funding amounts are rounded to the nearest thousands of dollars.

- 608 conference presentations
- 229 journal articles, and
- 345 invited colloquia.

Also encouraging is the number of colleges ($n = 1,116$) and secondary schools ($n = 565$) not funded in the NSF initiative that tried the materials developed by the projects during this time period in some or all of their calculus classes. An estimated 265 workshops and conferences to help the instructors at these institutions implement the materials were organized via the calculus initiative between 1988 and 1998.

F. NSF funding levels by year

Annual NSF funding per project ranged from $1,500 to $570,283, with an average per year of $111,699 and an average total for the life of a project of $186,458. Duration of awards was typically two years. The year with the most funded projects ($n = 37$) and the highest total budget ($4,074,942) was 1992, while the year with the highest funding per project was 1996, when $977,344 was divided among only three projects, for an average of $325,781 per project that year (see Table 9).[5]

Table 9 also reveals the periods of funding emphasis for specific objectives of reform. To illustrate, observe the NSF funding levels from 1988 to 1998 for projects stressing computer use. First, compare the number of projects funded and the average funding per project under calculus use to the corresponding total figures for each year. Although the numbers do not compare exactly, the general rises and falls in number of projects and average funding per project over this ten-year period is approximately equal for computer use and total annual figures. However, when making the same comparison between open-ended problems and total annual figures, very different trends are observed. Specifically, projects in 1988 that stressed open-ended problems were few in number ($n = 4$), but received funding levels much higher than the average for that year (roughly $82,000 as compared with $52,000). By 1993, projects stressing open-ended problems were still relatively few ($n = 9$), but their average funding had remained almost constant ($76,000), while the total annual average had increased significantly ($134,000). This indicates that the perceived importance of computer use in the mathematics community was closely aligned with national priorities for this objective, while this is not the case for open-ended problems. Similar comparisons can be made for each of the other national objectives.

General NSF funding strategies for the calculus initiative also can be observed in Table 9. For example, the total annual budget follows a relatively normal distribution, peaking in 1992 at approximately $4 million. The number of projects funded also follows this distribution until 1993, immediately after the peak funding year. At this time (and for the remainder of the initiative), NSF began to focus its support on fewer projects, implying higher average annual budgets per project (e.g., approximately $134,000 in 1993 and $148,000 in 1994). This shift in funding emphasis is even more apparent when examining the same figures for 1995–1997, the years when funding for new projects was no longer available. The dramatic decrease in number of funded projects and the corresponding increase in average annual budget per project reflects the fact that only a few projects ($n = 4$) were awarded funds that extended beyond the 1994 fiscal year. This change in funding strategy by NSF implies a decision to strongly support a few key projects in the final years of the initiative, a complete shift from the earlier strategy to award funds to as many promising projects as possible. Of course, this is a natural progression over the course of a multi-year initiative, as only select projects will be sustained while also establishing the widespread positive impact of their specific efforts.

---

[5] Although the last award in the calculus initiative was made in 1994, several projects were on-going, with the funding for the last project ending in 1997.

Student Achievement

One of the most important, yet often least investigated, outcomes of any educational innovation is the impact on student learning and achievement. A major goal of the calculus reform efforts is to create courses in which more students are able to obtain an understanding of the concepts and techniques that define calculus. In this chapter, an exploration of general research on student learning in calculus is presented, followed by a discussion of the known impact of calculus reform on student achievement.

It is important to note that "student achievement" is not a concept that is clearly defined nor which is generally agreed upon or even understood in the mathematics community. Discussions related to what a student should know and understand after completing a calculus course (whether that be one semester, a year, or two years) are heated and have yet to be resolved. Also being debated are which basic mathematical computations and skills should go along with the conceptual understanding. These are issues that never will be resolved completely (and likely shouldn't, if calculus is to remain a vital and "lively" course). Nonetheless, it is important to investigate student learning in calculus and collect data that will contribute to the knowledge base on how students learn calculus.

Data on student achievement also can make a powerful statement to college administrators and others who are skeptical about the changes being implemented in calculus courses. It is important to approach curricular change with the same scientific attitude used in disciplinary research; that is, it is only with an open mind and carefully constructed educational studies that mathematicians can discover the real impact of calculus reform (or any curricular change) on student learning. It is hard work—and will never yield a definitive answer to the question, "Are my students learning more calculus?" However, without the data, it will be impossible to determine a direction for change that will result in the greatest positive impact.

A. The literature on student learning in mathematics

A.1 *The notion of concept images*

Student achievement in mathematics is a very vague notion that is difficult to define. What does it mean to "understand" calculus? Is it even possible to describe the characteristics of a "good" calculus student? Hiebert and Carpenter (1992) define mathematical understanding as the development of a "mental representation" that is "part of a network of representations." The strength of the connections within the network determines the degree of understanding (p. 67). In other words, a mathematical concept is understood only if it can be related to other concepts that are already well understood by the individual (Dubinsky, 1999). This total cognitive structure is known as a *concept image* and includes mental pictures, as well as other associated properties and processes (Tall & Vinner, 1981).

Mathematical understanding as a concept image is well established and has been supported by many other mathematics educators, including Dewey (e.g., McLellan & Dewey, 1895) and Pólya (e.g., Pólya, 1957). Unfortunately, concept images are often confused with *concept definitions*, which are the formal words used by an individual to specify a concept (Tall & Vinner, 1981). In mathematics, this often translates into a list of rules and algorithms that guide the manipulation of a concept in problem solving. However, this is only part of the total concept image.

In many traditional forms of assessment, concept definitions are isolated from the total concept image and used to define student understanding of a mathematical idea. This narrow view of mathematical understanding often produces conflicts when a student's concept image and formal concept definition are not in agreement with each other or with the ones accepted by the mathematical community. An example of this conflict is seen in calculus when studying limits and continuity, notions which often produce concept images (based on the non-mathematical definitions of these words) which are not part of the formal theory as defined by mathematicians (Tall & Vinner, 1981). Therefore, it is important not only to address these conflicts in the context of a calculus course, but also to develop assessment instruments designed to measure overall concept images, as well as formal concept definitions.

It should also be recognized that student achievement is a function of time, with mental patterns and images constantly changing as a student's mathematical thinking matures (Dubinsky, 1992, 1999; Tall, 1977, 1978). This recognition implies that it is just as important to discover what a student does not understand as it is to discover what is understood, as this can help in the structuring of future activities that will help the student to build stronger mental images. It further implies that certain mathematical ideas, as well as various methods of delivery and assessment, may be appropriate at different stages of a student's mathematical development. This theory is, in part, the premise of calculus reform. Its implementation and subsequent impact on actual student achievement is the subject of this chapter.

A.2 *Student learning in calculus prior to 1986*

Efforts to improve calculus instruction did not originate in 1986 (see Figure 1). In fact, as recently as the 1960s there was a similar emphasis on changing the calculus course to better reflect the rapid developments in science and the needs of industry (see Chapter 1 for a discussion of this history). Unfortunately, evaluations of the impact of these alternative environments on student learning were sparse and mostly in the form of doctoral dissertations (e.g., Levine, 1968; McKeen, 1970; Monroe, 1966; Shelton, 1965). Nonetheless, the similarities between these studies and the current reform efforts are striking. For example, three distinct studies investigated the impact of cooperative learning on student achievement in calculus with one finding in favor of the traditional students (Larsen, 1961), and two revealing no significant differences between groups (Cummins, 1960; Davidson, 1970) except in the area of conceptual understanding (Cummins, 1960). The effect of computer programming on overall student achievement is reported in six studies, with no significant effect (Bell, 1970; Buck, 1962; De Boar, 1974; Fiedler, 1969; Holoien, 1971; May, 1964). When conceptual understanding was tested independent of manipulative skills, the computer students performed significantly higher than the traditional students (Bell, 1970). The use of computers in calculus to assist in extra programmed learning (Stannard, 1966) or as a general supplementary tool (Smith, 1970) also was found to have a positive effect on student achievement.

The small number of evaluation studies on the calculus reform efforts of the 1960s and the de-emphasis on calculus reform for almost two decades after this time did not completely eliminate interest in calculus. In fact, much relevant research on student learning in calculus conducted during this time would pave the way for the current reform efforts. These studies can and should be used to guide the interpretation of results from the evaluation of reform efforts since 1986. Specifically, information on student understanding of functions, limits and continuity, derivatives, and integrals, as well as the influence of technology and cooperative learning on calculus before the current reform efforts, are discussed in this section.

In general, the literature suggests that the mathematical understanding of most calculus students on fundamental topics such as those listed is limited to formulas, algorithms, and basic techniques (Cipra, 1988; Ferrini-Mundy & Graham, 1991; Frid, 1994; Ganter, 1997a,b, 1999; McCallum, 2000; NSF 1987, 1991; Peterson, 1987; Smith, 2000). Students can easily respond to standard questions posed in a predictable way, but when pushed to reveal any deeper understanding of the concepts, they often have difficulties (Selden, et al., 1989; Tall, 1990). This minimal understanding is often further weakened by prior notions of these ideas that are in conflict with the formal mathematical definitions, as was previously discussed for limits and continuity (see section A.1; Confrey, 1980; Tall and Schwarzenberger, 1978). Prior experiences in mathematics can also contribute to firmly believed misconceptions about functions, such as the notion that a function is a collection of points, that it must be linear, and that it can only be represented as a formula (Dreyfus & Eisenberg, 1982, 1984; Graham & Ferrini-Mundy, 1989; Monk, 1989). These weaknesses and misconceptions often lead to the inability of students in calculus to think beyond the most routine aspects of differentiation and integration (Orton, 1983a,b). It is believed by some that an intuitive understanding of these concepts must therefore be developed through explorations using alternative representations and the reinforcement of ideas developed cooperatively by secondary school, college, and university faculty.

An interesting idea explored in the teaching of calculus in the 1970s was self-pacing. Self-paced courses use individualized student learning in combination with tests for mastery of material and lecture to motivate learning, rather than impart knowledge (Teles, 1992). A number of studies on the impact of this method found significant differences in student achievement favoring the self-paced students (Jackson, 1979; Pascarella, 1978; Peluso & Baranchik, 1977; Struik & Flexer, 1977; Taylor, 1977), with no significant differences in a few others (Klopfenstein, 1977; Moore, 1976; Thompson, 1979). The concept of a mathematics learning center evolved from the self-paced courses of the 1970s; few institutions continue to offer this as a regular option for calculus students (Teles, 1992).

Multiple prior studies also exist that support some of the alternative methods of presentation that have been embraced by the recent reform efforts in calculus. These studies have investigated the impact of activities such as cooperative learning (Brechting & Hirsch, 1977; Davidson, 1985, 1990; Hurley, 1983; Kroll, 1989; Loomer, 1976; Lovelace & McKnight, 1980; Slavin, 1983, 1985; Urion & Davidson, 1992) and the use of technology (Basil, 1974; Flores, 1985; Hamm, 1990; Hawker, 1986; Heid, 1984, 1988; Johnson, et al., 1986; Murphy, 1975; Palmiter, 1986,1991; Schrock, 1989; Tall, 1990; Tall, et al., 1984; Tufte, 1990; Webb, et al., 1986; Yackel, et al., 1991), concluding that they do in fact enhance student learning in mathematics in most cases. However, only a handful of these studies address the impact of these activities specifically in a calculus course. Technology is reportedly affecting over 90% of the reform efforts (see Table 6) and is "the most visible force for change in the mathematics curriculum." (Steen, 1987, p.3) Cooperative learning is being used as a pedagogical strategy in combination with technology, as well as in its own right. Many other components of reform, such as technical writing, long-term projects, and alternative assessment are also having an effect on student achievement. A better understanding of the impact of these various components is clearly needed to make informed decisions about the future of calculus reform.

B. The impact of current reform efforts

For the purpose of this study, student achievement is defined to be results from studies designed to measure the performance of students on problems involving the mathematical content of the calculus course. One of many ways this can be studied is through the relative performance of reform and traditional students on common exams or problem solving interviews. Achievement of reform students can also be measured in a non-comparative manner by analyzing, for example, writing samples from student projects or journals. The analysis of NSF files revealed that 34 of the 127 funded projects reported results from the collection of such data on student achievement (see Table 2). Fifty-nine additional studies on student achievement that were not funded by NSF were also reviewed. These additional studies included 11 doctoral dissertations and one masters thesis (see Table 10 for a summary of the studies analyzed in this chapter). Overall, 98 of the 111 studies reviewed (88%) report that the impact of reform efforts on at least one measure of student achievement is positive. Seventeen of the studies concluded that the reform efforts had a negative impact in at least one area (usually computational skills). The information reviewed also strongly suggests that the particular reform implemented, as well as the nature of the students and instructors in the reform classes, has a great

Table 10. *Sources of evaluation data for student achievement by Carnegie classification* and NSF funding.***

| Source*** | Frequency (n = 111) | | | | | | | | | | | |
| | Research/doctoral | | Comprehensive | | Liberal Arts | | Two-Year | | High School | | Total | |
	NSF	Non-NSF	NSF	Non-NSF	NSF	Non-NSF	NSF	Non-NSF	NSF	Non-NSF	NSF	Non-NSF
Published article	9	8	6	14	0	1	0	4	0	2	15	29
Published report/ collections of articles	0	3	0	1	0	0	0	0	0	0	0	4
Conference proceedings	2	2	10	4	0	1	1	0	0	0	13	7
Unpublished manuscript	0	2	2	1	2	0	2	0	0	0	6	3
Doctoral dissertation	5	11	0	0	0	0	0	0	0	0	5	11
Masters thesis	0	0	1	1	0	0	0	0	0	0	1	1
Internal institutional report or report to NSF	2	0	2	1	0	0	1	0	0	0	5	1
Individual letter	0	1	2	0	2	1	3	1	0	0	7	3
Total	**18**	**27**	**23**	**22**	**4**	**3**	**7**	**5**	**0**	**2**	**52**	**59**

*Indicates the Carnegie classification for the primary institution where the work was conducted.
**Indicates whether the document was produced as part of a NSF-funded project.
***For complete citations on all references used in the compilation of data for this project, see the MAA website (www.maa.org).

effect on student outcomes. Specifically, certain trends were observed in projects that used technology and group work. A number of institutions also came to the same conclusions when monitoring the success of their reform students upon returning to more traditional subsequent courses. These results, as well as studies on the effect of several specific reform curricula, are discussed in the following sections.

B.1 *The impact of technology*

The technology used in calculus reform efforts has taken two forms: computers and graphing calculators. Because most projects have chosen to use one or the other (not both), it is possible to discuss the relative impact of each of these technologies as studied in the projects. The selection of computers vs. graphing calculators varied by institution type (see Table 7). Sixty-eight distinct sources of information were used in the compilation of the data reported here (see Table 11).

Computers. Seventeen NSF-funded projects and eight other projects reported results on student achievement as impacted by computer technology. These studies included results from more than 14 research universities, seven comprehensive universities, two liberal arts institutions, and four two-year colleges.[6] Nine of the research institutions reported increases in conceptual understanding, greater facility with visualization and graphical representations, and the ability to solve a wider variety of more difficult problems, without any loss of computational skills. The other research institutions reported questionable or lower levels of computational skills after completing a calculus course using computers as compared with the students in the traditional course (Aspinwall, 1994; Ganter & Jiroutek, 2000; Jackson, 1996; Padgett, 1994; Soto-Johnson, 1996). The studies from comprehensive, liberal arts, and two-year colleges have more mixed results (Allen, 1995; Jackson, 1996; Pustejovsky, 1996; Soto-Johnson, 1996; Webber, 1996). Some institutions commented that technology, if not used appropriately, can overwhelm the goal of reform efforts to

---

[6] Numbers reported for institution types throughout this report are not exact due to the fact that a number of studies did not report the name of the institution being studied (e.g., those submitted as manuscripts for publication). Therefore, numbers reported for institution types are based only on studies that provided this information.

improve student understanding (Jackson, 1996; Padgett, 1994; Rochowicz, 1993). And although students in courses using computers often have a better understanding of continuity, differentiation, and integration, they consistently score lower than traditional students on problems involving limits (Allen, 1995; Dubinsky, 1999; Francis, 1992; Ganter & Jiroutek, 2000; Simonsen, 1995). The most common comment from all studies on the impact of technology is that, although it is uncertain whether the use of computers has improved student performance, the investigators are fairly certain that the students in the reform classes have done at least as well as those in the traditional classes.

Computers can help more students succeed in calculus, with at least three institutions observing a remarkable reduction in the number of students failing the course. Significant differences in the number of students receiving A's and B's were also reported, especially after the instructor has had several semesters of experience using computers in the course (Donaldson, 1993; Monteferrante, 1993; Silverberg, 1994). A comment frequently made by instructors is that even students who are receiving C's in the reform sections have a much deeper understanding of the concepts than former students who received equivalent or higher grades in their traditional course.

Graphing calculators. Positive student achievement outcomes were also reported when using graphing calculators by more than 15 research universities, five comprehensive universities, two liberal arts colleges, and 11 two-year colleges. The results were very similar to those when using computers, as virtually all of the research institutions reported greater conceptual understanding with equivalent or lower computational skills. One research institution that had computer, graphing calculator, and traditional offerings reported that the computer sections scored higher on course grades and other measures in some subsequent courses. In other subsequent courses their performance was indistinguishable from that of students who had been enrolled in the graphing calculator and traditional calculus sections. The other types of institutions were again very mixed in their results, with less conclusive studies that point only to the possibility of improved performance. Failure and withdrawal rates were again reported to be significantly lower at several institutions (e.g., Schneider, 1995).

One consistent result among students using graphing calculators is the significant difference when compared with traditional students in their ability to interpret graphs and to discuss the relationship between the

Table 11. *Sources of evaluation data for student achievement when using technology by Carnegie classification* and NSF funding.***

Source***	Research/doctoral NSF	Research/doctoral Non-NSF	Comprehensive NSF	Comprehensive Non-NSF	Liberal Arts NSF	Liberal Arts Non-NSF	Two-Year NSF	Two-Year Non-NSF	High School NSF	High School Non-NSF	Total NSF	Total Non-NSF
					Frequency (*n* = 68)							
Published article	6	4	2	4	0	0	0	1	0	0	**8**	**9**
Published report/ collections of articles	0	2	0	1	0	0	0	0	0	0	**0**	**3**
Conference proceedings	1	2	4	4	0	1	1	1	0	1	**6**	**9**
Unpublished manuscript	0	2	3	1	2	0	2	0	0	0	**7**	**3**
Doctoral dissertation	2	5	0	0	0	0	0	0	0	0	**2**	**5**
Masters thesis	0	0	0	1	0	0	0	0	0	0	**0**	**1**
Internal institutional report or report to NSF	1	0	0	0	1	0	1	0	0	0	**3**	**0**
Individual letter	1	1	2	0	4	1	1	0	0	2	**8**	**4**
Total	**11**	**16**	**11**	**11**	**7**	**2**	**5**	**2**	**0**	**3**	**34**	**34**

* Indicates the Carnegie classification for the primary institution where the work was conducted.
** Indicates whether the document was produced as part of a NSF-funded project.
*** For complete citations on all references used in the compilation of data for this project, see the MAA website (www.maa.org).

graph of a function and that of its derivative and integral (Alexander, 1997; Barton, 1995; Brunett, 1996; Hershberger & Plantholt, 1994; Pence, 1996; Penn, 1994). However, while the graphical nature of the functions emphasized by graphing calculators helps many students, it can be a difficult obstacle for students who have previously excelled because of their strong computational skills. Clearly, individual students show definite preferences for certain representations, with a wide variety of factors influencing these preferences (Hart, 1992). The ability to recognize function types both graphically and analytically was also enhanced by use of the graphing calculator, although both graphing calculator and traditional students were equally weak in their understanding of the definition of a function (Hart, 1992; Williams, 1994).

Students' grades in the course are not significantly different in most cases between the traditional and graphing calculator courses for those completing the course. However, several institutions report that significantly higher numbers of traditional students are withdrawing from or failing calculus than those in courses with graphing calculators (Carruthers, 1996; Pence, 1996; Ratay, 1992, 1993, 1994; Waggoner, 1996). This occurs in spite of the fact that the students with strong mathematics backgrounds (potential mathematics majors) are often encouraged to take the traditional course at these institutions. An exception is for students who enroll in and successfully complete a calculus course with graphing calculators and then enroll in a traditional subsequent calculus course. In this situation, a noticeable decline in the performance of the student as measured in the traditional context can occur, although this is not always the case (Armstrong, Garner, & Wynn, 1994). It should also be noted that results from studies using a common computational exam (instead of course grades) to compare achievement were less conclusive, with about 50% resulting in equal performance between the reform and traditional groups and another 45% concluding that the traditional students had significantly higher scores (e.g., Schneider, 1995; Williams, 1994). Very few studies, therefore, resulted in higher computational scores for the reform students. Interestingly, very little formalized testing has been conducted to compare areas of achievement stressed by reform courses, such as conceptual understanding, critical thinking, or technical writing.

B.2 *The impact of projects and group work*

Another area investigated by several institutions was the effect of long-term projects and group work on student performance. Twenty-eight studies (with information from 80 institutions) reported specific efforts to investigate the effect of these alternative pedagogical styles on student performance. A number of patterns emerged from these reports, including the type of student that is likely to excel in environments using these techniques. For example, evidence suggests that "above average" students in mathematics, students who do not perform well on traditional tests, and engineering majors tend to have the most success with long-term projects. Group work contributed to greater success in calculus[7] for a wider student population, especially students whose prior mathematical performance was average or below average (Donaldson, 1993; Monteferrante, 1993; Silverberg, 1993). Not surprisingly, the available information suggests that both pedagogical methods positively impact students' ability to write about and explain mathematics (Cribbs, et al., 1996; Hershberger & Plantholt, 1994; Pustejovsky, 1996). In fact, there is evidence that group work encourages students to attempt more difficult problems using multiple problem-solving strategies, even when they do not yet fully understand the mathematics necessary to complete the work. This situation often leads to self-discovery of mathematical concepts by the student groups (Crocker, 1991).

Projects and group work also have an impact on grade distribution, although not in consistent ways. For example, one research institution reported that projects made the final grade distribution more bi-polar, with very few "C" students, while another research institution reported that projects were "the great equalizer," with more C's awarded due to the subjective nature of grading the projects and therefore an inability to justify grades at the extremes. The organization of projects as individual or group assignments also impacts

---

[7] For the purpose of this report, "success" in calculus is defined to be the completion of the course with a grade that enables them to enroll in the next level of mathematics or a passing grade on an examination used for the study. The instrument used to measure this success varies for the different studies analyzed.

course grade distributions, as group work tends to "equalize" the grade distribution more than assignments graded on an individual basis.

The evidence presented here suggests that long-term projects and group work, which are often used in combination in reform efforts, each target very different groups of students. This result would imply that the way in which these two reform strategies are combined and implemented can directly affect the student populations that will succeed. The seemingly random impact on grade distribution confirms this interpretation, as it is likely the result of two methods that naturally "team up" but that in fact may serve two entirely different pedagogical needs. Extensive work on the interaction of projects and group work is therefore needed to fully understand the ways in which they may be most effectively utilized.

B.3 *Compatibility of reform and traditional courses*

An area of concern for many institutions implementing reform strategies is the ability of students to successfully transfer from one type of course to another for the various terms of calculus. For example, can students who have a reform course with computers for differential calculus switch to a traditional course for integral calculus and still perform at the same level as those students who were enrolled in the traditional course for both terms? As expected, the results vary and are dependent in large part on the types of assessment measures used in each of the courses. It appears that the degree to which the reform course differs from the traditional one is the most influential component of their interchangeability. Students from reform courses with a very heavy emphasis on computers and applied problems, two major components of calculus reform, typically have more difficulty when switching to a traditional course than those students from reform courses that use computer lab activities infrequently and in conjunction with a traditional or "conservative" reform textbook. Students also have less difficulty making the switch from a graphing calculator environment to a traditional one, probably because the calculator is often used as a supplement to the traditional environment and also can be transported to the traditional course after the switch is made. This result can vary with the level of implementation, as previously discussed.

C. The impact of specific projects

Although it is not the intent of this report to evaluate the efforts of specific institutions, several projects supported by NSF have been widely disseminated and therefore it seems appropriate to discuss the impact of those for which information is available from multiple institutions. The efforts discussed here were chosen solely on the basis of information submitted by dissemination sites not formally affiliated with the original development of the given strategy. No conclusions about the merit of the listed projects or their relative importance should be inferred, as this report is only intended to be a synthesis of relevant information, not an attempt to judge the value of individual efforts.

C.1 *The Calculus Consortium based at Harvard University*

The work of the Calculus Consortium based at Harvard University (CCH) is the most widely disseminated of the curriculum development projects funded by NSF. Therefore, the results reported here naturally support many of the findings previously discussed for all studies of student achievement. Included in this review are studies and discussions of achievement resulting from the implementation of the CCH materials at 17 institutions. The institutions included seven research universities, three comprehensive universities, one liberal arts institution, and six two-year colleges (see Table 12) and are fairly representative of the entire population of known CCH users at the postsecondary level.[8] The studies included, among others, two doc-

---

[8] Based on the 1993 adoption list for the single-variable calculus textbook written as part of the CCH project and published by John Wiley & Sons.

Table 12. *Distribution by Carnegie classification of evaluation studies conducted at other implementation sites for select NSF-funded projects.*

Project	Frequency				
	Research/doctoral	Comprehensive	Liberal arts	Two-year	Total
Calculus Consortium based at Harvard University	7	3	1	6	17
Calculus, Concepts, Computers, and Cooperative Learning (C⁴L)	1	2	1	0	4
Calculus and Mathematica™	3*	0	0	0	3
Project CALC	1*	0	1	0	2
Calculators in the Calculus Curriculum	3	0	0	0	3
Total	**15**	**5**	**3**	**6**	**29**

*One site that implemented and evaluated a combination of the Calculus and Mathematica program with Project CALC is recorded for both projects.

toral dissertations and one masters thesis; longitudinal studies[9] were conducted at several of the institutions. Eleven of the institutions explicitly mentioned that technology was being used in combination with the CCH materials. Specifically, ten institutions examined the impact of the CCH materials in combination with graphing calculators, the most common mode of technology reported in combination with the CCH curriculum.

Students in calculus courses utilizing the CCH materials score significantly higher than their traditional peers on graphically-oriented problems (Allen, 1995; Brunett, 1996; Penn, 1994; Schneider, 1995; Tidmore, 1994). This includes both problems that ask about data presented graphically and ones that ask the student to produce graphs related to the given graph; e.g., the graphs of the derivative and the antiderivative. CCH students also fare better on problems that ask for explanations and those that require students to use or interpret multiple representations of the same concept (Alexander, 1997; Allen, 1995; Brunett, 1996; Cribbs, et al., 1996; Hershberger & Plantholt, 1994; Tidmore, 1994; Williams, 1994).

The studies analyzed here report that, in general, course grades are not affected by the use of the CCH materials for students successfully completing one year of calculus (Cribbs, et al., 1996; Lepowsky, 1996; Pence, 1996; Prettyman, 1996; Schneider, 1995). However, several studies contrasting CCH and traditional students report that significantly fewer CCH students are failing or withdrawing from the course (Carruthers, 1996; Pence, 1996; Ratay, 1994; Waggoner, 1996). An additional finding of interest that reflects the overall results previously discussed is that a significant drop in the course grades of the CCH students in Calculus I is observed when they transfer to a traditional Calculus II course (Armstrong, Garner, and Wynn, 1994; Shah, 1996).

Also supporting the overall results is the unclear effect of the CCH curriculum on students' ability to complete traditional computational problems. Results on common computational exams for CCH and traditional students yielded mixed results, with five studies favoring the CCH students, four studies favoring the traditional students, and three studies showing no difference between the two groups. Again, although more studies are needed to understand these discrepancies, it would appear that the performance of the CCH students on traditional computational measures has not been altered significantly when compared to the performance of the students in traditional courses on these same measures.

---

[9] Longitudinal studies are defined to be ones that collected data for at least one year.

C.2 *Calculus, Concepts, Computers, and Cooperative Learning*

The Calculus, Concepts, Computers, and Cooperative Learning (C^4L) project, originating at Purdue University, places a heavy emphasis on research in the area of student learning and the use of that research to influence classroom practice. Therefore, a number of studies on student achievement as affected by the C^4L curriculum have been conducted. Four of these studies, conducted by individuals not directly affiliated with the C^4L project, were reviewed.

The institutions at which the studies were completed include one research university, two comprehensive universities, and one liberal arts college (see Table 12). These studies focused mostly on the impact of C^4L on course grades and, more specifically, on the level of understanding for students receiving various course grades as compared to traditional students receiving the same grade. Two of the studies, which utilized a very similar rigorous methodology, concluded that the C^4L curriculum helped more students succeed in the calculus course. A remarkable reduction in the number of students receiving grades below C was observed, and students who had historically been C students were excelling in this environment, implying that the C^4L curriculum has the greatest positive impact on students who have not done well in the traditional environment. Informal observations at a third institution confirmed these results (Donaldson, 1993; Monteferrante, 1993; Silverberg, 1993). As with the overall results, common traditional exams administered at two institutions yielded mixed results, with data being inconclusive or in favor of the C^4L students (Silverberg, 1993; Webber, 1996).

C.3 *Calculus and Mathematica™* and *Project CALC*

Calculus and Mathematica™ was conceived at the University of Illinois at Urbana-Champaign and is arguably one of the most "radical" of the NSF-funded reform projects. It incorporates many of the elements of reform, such as cooperative learning, project work, and a discovery approach. However, Calculus and Mathematica™, as the title of the project implies, places great emphasis on harnessing the power and utility of the computer in every facet of the course. This curriculum uses the computer not only as a tool for understanding calculus concepts, but also as an organizational mechanism for the course. Students are thrust into a "paperless" environment, with readings, assignments, and activities all completed on the computer. This implies that the implementation of Calculus and Mathematica™ is necessarily limited to institutions that have the computational power to handle not only Mathematica™, but also the enormous amount of computer materials generated by the course. Students at these institutions appear to benefit in many ways, as seen in two dissertation studies conducted at The Ohio State University, one of the implementation sites for Calculus and Mathematica™ (see Table 12). According to one study, Calculus and Mathematica™ students learned to be very independent thinkers who responded positively to a cooperative setting and had confidence in their ability as problem solvers, whether on or off the computer (Crocker, 1991). Students became more willing over time to attempt one problem using multiple approaches and could make mathematical connections in a conceptual way prior to a formal understanding of underlying principles. The other study demonstrated that they were also better able to make connections between graphical and symbolic representations than students in traditional or graphing calculator courses (Porzio, 1994).

Project CALC, originating at Duke University, has received attention for its unique emphasis on cooperative learning in combination with computer lab activities. Although similar in this way to Calculus and Mathematica™, it differs in that each of these entities plays an equally important role in the course. Project CALC has also served as a leader in the evaluation of reform activities, collecting data on the impact of their efforts since the formal beginning of the project in 1988 (see Bookman & Friedman, 1994, 1998). Various components of the evaluation model developed by the project team have been used in the evaluation of reform efforts at other institutions, many of which were not using Project CALC materials. One study that investigated the impact of Project CALC at another institution confirmed the results of the Calculus and Mathematica™ study; that is, that students in a cooperative setting with computer technology develop a strong ability to process and apply mathematical ideas, with or without a computer (Pustejovsky, 1996).

The similarities between the two projects are further emphasized by a dissertation study comparing the achievement of Calculus II students from Project CALC, Calculus and Mathematica™, and traditional classes at different universities. The achievement measure, which utilized several questions developed by the Project CALC team, showed that there were no significant differences in the conceptual understanding of infinite series for students in courses using these three methods (Soto-Johnson, 1996). General computational scores for the traditional students were higher than those for the reform students. Students' attitudes toward the importance of technology in solving mathematical problems and toward mathematics in general were significantly higher for students in the two reform courses than those in the traditional course. And, classroom observations yielded the highest rate of student-student interaction in the Project CALC course, with Calculus and Mathematica™ second, followed by the traditional course, in which very little student-student interaction was observed. Not surprisingly then, it can be inferred from these combined results that students will excel in their performance and attitudes on those components of the calculus course—both conceptual and pedagogical—that are most heavily emphasized.

C.4 *Calculators in the Calculus Curriculum*

Calculators in the Calculus Curriculum, originating at Oregon State University, was developed in cooperation with Hewlitt Packard and emphasizes the use of graphing calculators in the calculus course. The project has also been very active in the evaluation of their efforts and the impact of graphing calculators on student learning through a series of doctoral dissertations conducted on-site. Three of these studies were complete and available for the writing of this report; their results are reported here (see Table 12).

A comparative study between the graphing calculator students and those in traditional courses indicates that the reform students have greater facility with graphical and numerical representations (Hart, 1992), as was also concluded in several previously discussed studies. The calculator students are also less likely to compartmentalize knowledge or use single representations when solving problems, making the use of this technology more difficult for students who rely heavily on their strong computational skills. Graphing calculators can be used successfully for problem-solving activities in both student-centered and instructor-centered environments (Utter, 1996). An added benefit when used in combination with a student-centered environment is the stronger ability demonstrated by these students to interpret graphs and tables and to better explain the concepts of differential calculus.

Familiarizing teachers with research that shows the effect of various learning environments is a necessary and powerful tool in influencing their ability to change their approach in the classroom. For example, one study concluded that student errors in explaining the concept of derivative in an environment using graphing calculators is often simply a misunderstanding or misuse of language (Zandieh, 1997). This implies that technical writing and group discussions of concepts are also very important if students are to confront and correct these misunderstandings at an early stage of their learning in calculus.

D. Conclusions

Clearly, student achievement has been impacted by the calculus reform efforts. What is perhaps less clear is the degree to which achievement has been affected and the appropriate "mixture" of reform ideas that should be implemented at various institutions to achieve the greatest positive impact. The available information indicates that the use of technology, long-term projects, or group work—or any combination of these—results in students who have better problem solving skills and a stronger conceptual understanding of the major ideas in calculus. An emphasis on visual and physical calculus with strong connections to the corresponding symbolic representations has also been shown to enhance student understandings of calculus (Ahmadi, 1995; Ellis, 2000; Frid, 1994; Wells, 1995).

The effect of reform on more traditional computational skills is still uncertain because of the small amount of research in this area, as well as the conflicting results from the research that is available. Even less studied is the impact of reform methodologies on student skills that are difficult to quantify, such as

- the ability to communicate mathematics
- an understanding of when and how to apply mathematics in real problems, and
- the effectiveness of students when working in teams.

The controversy surrounding these skills and their relative importance when compared with other possible learning outcomes as goals for the calculus course implies a strong need for further investigation. Extensive work on the interaction of projects and group work—as well as other teaching tools and methodologies—is needed to fully understand the ways in which they may be most effectively utilized.

Student Attitudes

With the exception of improving student achievement (see Chapter 4), changing student attitudes about the importance and relevance of mathematics was considered the primary goal for many of the calculus reform projects. It was generally believed by mathematicians in the mid-1980s that many students regarded calculus (and mathematics in general) as a necessary evil in their quest to earn an undergraduate degree. In fact, many college faculty (including mathematicians) readily agreed that calculus was simply a barrier to be crossed by students wanting to pursue careers in science or engineering. And, to compound the problem, it was (and unfortunately still is) socially acceptable—even fashionable—to dislike mathematics.

So, why is this a problem? As is discussed in this chapter, student attitudes about mathematics directly affect their ability and decision to complete the calculus courses in which they enroll. Attitudes can also impact the number of students electing to enroll in subsequent mathematics courses—especially if they are not required in their major area of study. The bottom line: students who do not understand the relevance of mathematics to their education and future will not believe it is important to learn or continue to study mathematics.

There is a self-serving interest as well for the mathematics community. Specifically, undergraduate enrollments in mathematics departments (math majors) are dropping at most institutions (Loftsgaarden, Rung, & Watkins, 1997; NSB, 1998). Why? Among other reasons, students want to pursue an undergraduate degree that will result in what they perceive as a desirable (and, often, lucrative) career. If students are not exposed to the power of mathematics not just for the sake of mathematics, but also as a field of study and preparation for a suitable career, they simply will not choose to become math majors in many cases. Not to mention that many students taking calculus (freshmen) are undecided about their careers. This makes calculus an ideal course in which to excite students about studying mathematics.

Therefore, understanding what students think about calculus courses and the impact that calculus reform has had on those attitudes is important not only for the calculus course, but also for the health and future of many mathematics departments. This chapter explores the knowledge base about student attitudes in calculus prior to the current period of reform, followed by discussions about the impact of the reform efforts on student attitudes. Several causes for student resistance to reform as identified by the studies also are discussed in this chapter.

A. Student attitudes in calculus prior to 1986

Investigations of student attitudes about calculus prior to the current reform efforts are scarce. Evidence from such studies that exist, as well as from the experiences of many individual mathematics instructors, indicates that students did not regard calculus as a powerful and necessary tool in the study of scientific problems. On the contrary, they simply regarded calculus as a hurdle that they must clear before continuing with the courses they would really need in their future endeavors. Many faculty also regarded calculus as a "filter," rather than a "pump," for courses in science, mathematics and engineering (Steen, 1987; White, 1987). It was understood—and generally considered to be acceptable—that calculus was the mathematics course most dreaded by thousands of students across the country (Kolata, 1987). This attitude only served to fuel the negative views of the course held by students.

Some studies on the impact of alternative pedagogical methods on student attitudes toward calculus were conducted before 1986. For example, the effect of computers was investigated with mixed results. Some studies revealed no differences between the attitudes of these students and those in traditional courses (Hamm, 1990; Hawker, 1986), while others concluded that these students enjoyed mathematics more (Judson, 1988) and had more confidence in their own abilities (Schrock, 1989).

The use of self-paced courses in calculus, which yielded impressive student achievement results, yielded significantly lower satisfaction rates than the traditional courses (see also Chapter 4, Section A.2). Although the personal attention of such courses was appealing to students, they complained that the course was very time consuming and imposed too much responsibility on the students, making it easy to procrastinate (Teles, 1992). Therefore, the interest in developing self-paced courses declined rapidly after the 1970s.

A few studies investigating the effect of large lectures and televised instruction were also conducted prior to 1986. The somewhat radical change from the small class setting of the 1950s to the large lectures and televised distance learning courses of the 1960s was developed in response to the increasing demand for courses and the consequential shortage of qualified instructors for these courses (see Figure 1)[10]. In spite of this dramatic change in the classroom environment, very little research was conducted to investigate the impact of these large lectures on student learning and attitudes. Of the several studies available (Cope, 1980; Jackson, 1979; May, 1962; Moise, 1965; Stockton, 1960; Turner, et al., 1966), only one demonstrated any significant difference in student achievement (Cope, 1980). One study on the use of televised lectures did reveal very negative student attitudes about the method (Dyer-Bennett, et al., 1958). In fact, almost all of the students felt that the class had not held their interest and that the opportunity to discuss problems and ask questions was critical to success in a calculus course. Of course, these concerns could also be relevant in a non-televised large lecture course.

These and other reactions to calculus courses of the several decades prior to 1986 led in part to the reform efforts of today. However, the lack of studies to document the impact of such efforts on student attitudes is just as real in the current environment as it was then. And, the need for such information is just as vital to the future of these efforts. Therefore, the remainder of this chapter will discuss the issues associated with student attitudes toward calculus and the results from evaluations of those attitudes as affected by current work in calculus reform.

B. The impact of current reform efforts

Student reactions to reform calculus were monitored by a number of NSF-funded projects, including 29 research universities, seven comprehensive universities, nine liberal arts colleges, and seven two-year colleges. Information on changes in student attitudes as affected by calculus reform also was received from 28 other institutions, for a total of 80 institutions studied. Distinct sources in which this information is reported number 65, including three doctoral dissertations and two masters theses (see Table 13). Additional information on student attitudes about calculus reform was found in a report from the Course and Curriculum Development (CCD) program evaluation at NSF, published in 1996 (see Eiseman, et al., 1996). It is the synthesis of this wealth of information that is reported here.

Student attitudes about reform vary dramatically within and between institutions, depending on the type of reform, level of implementation, attitude of instructor, degree of help available as perceived by the students, and amount of experience with the reform method (for both the student and the instructor). However,

---

[10] Of course, the financial savings to institutions was also a benefit of this method and ultimately has been the justification for the continuation of the large lecture format long after the shortage of qualified mathematicians has been corrected.

Table 13. *Sources of evaluation data for student attitudes by Carnegie classification* and NSF funding.***

Source***	Research/doctoral		Comprehensive		Liberal Arts		Two-Year		High School		Total	
	NSF	Non-NSF	NSF	Non-NSF	NSF	Non-NSF	NSF	Non-NSF	NSF	Non-NSF	NSF	Non-NSF
Published article	9	0	5	7	0	2	1	3	0	0	15	12
Published report/ collections of articles	0	3	0	1	0	0	0	0	0	0	0	4
Conference proceedings	1	0	6	1	0	0	1	1	0	1	8	3
Unpublished manuscript	0	1	2	1	2	0	2	0	0	0	6	2
Doctoral dissertation	0	3	0	0	0	0	0	0	0	0	0	3
Masters thesis	0	1	1	0	0	0	0	0	0	0	1	1
Internal institutional report or report to NSF	1	0	0	1	0	0	1	0	0	0	2	1
Individual letter	0	1	2	0	2	1	1	0	0	0	5	2
Total	**11**	**9**	**16**	**11**	**4**	**3**	**6**	**4**	**0**	**1**	**37**	**28**

* Indicates the Carnegie classification for the primary institution where the work was conducted.
** Indicates whether the document was produced as part of a NSF-funded project.
*** For complete citations on all references used in the compilation of data for this project, see the MAA website (www.maa.org).

there are several trends in student attitudes that emerge from the information. Perhaps the most apparent is that students are not ambivalent about reform; they either love it or they hate it. In general, when excluding other factors, the students who respond most positively to reform methods fall into one or more of the following categories:

- little or no prior experience with calculus
- moderate level of success in traditional mathematics courses
- motivated to learn
- poor test takers
- women
- minorities
- third- or fourth-year students who had reform calculus in their first year.

On the other hand, the students most negative about reform often have one or more of the following traits:

- prior experience with calculus, usually in the form of a high school Advanced Placement (AP) course
- high levels of success in traditional mathematics courses
- enrolled at institution that offers both traditional and reform courses.

Of course, these generalizations are not true for all students at all of the institutions implementing reform methods. A sufficient number of studies report some or all of these trends to warrant further discussion.

B.1 *Experiences in mathematics*

As stated above, prior mathematical experiences can greatly affect a student's attitude about reform in calculus. In particular, the biggest opponents to reform efforts are usually students who have been very successful in mathematics, their high school career culminating with a year of calculus but no AP credit awarded. Students in this particular group often complain that reform classes are "not mathematics" and that they "have had calculus and know what they should be learning." They are upset because the instructor will not simply "let them use the formulas" and are afraid that they will be at a disadvantage in future courses because they have not been doing "real calculus." (Eiseman, et al., 1996; Ganter & Jiroutek, 2000; Pustejovsky, 1996; Salem, et al., 1996)

They also complain that reform calculus is more difficult (meaning more time consuming) than the traditional course they had in high school. Or, on the contrary, they sometimes complain that it is less difficult (Eiseman, et al., 1996). Interestingly, these two conflicting results about the difficulty of reform came from institutions using the same reform curriculum, highlighting the fact that even students with similar high school experiences who are enrolled in similar reform calculus courses at the college level may still walk away with different perceptions of the course. Whether those perceptions are real or imagined has yet to be determined.

The negative responses of these seemingly successful mathematics students toward reform may simply be a reflection of the positive experiences that most of these students have had with traditional mathematics. It is not surprising that they would resist a course that is so different from the courses with which they have had so much success. Conversely, this is the very reason that many students with a less successful mathematical history (which is often the majority of the students in a calculus course) react positively to reform and are more likely to excel in this environment. Specifically, the alternative methods of assessment that often accompany reform efforts (e.g., term papers from projects and computer lab reports) mean that students with a variety of strengths—not just good test takers—now have a better opportunity to succeed in calculus. Of course, the appropriate balance between testing and other measures is still critical to fully assess a student's understanding of calculus.

B.2 *Special groups*

A number of studies noted that women and minorities tend to respond favorably to certain aspects of reform. In particular, four studies concluded that reform efforts using cooperative learning and group activities yield especially positive results for women and minorities. Specifically, most women and minority students preferred courses that incorporated small workshop sessions for completion of cooperative activities, in addition to regular class meetings, over the more traditional format (Bonsangue, 1994; Hastings, 1997). The students noted that the accountability of the workshops motivated them to study. Even the students who dropped the course in favor of the more traditional model cited an unmanageable workload, not a dissatisfaction with the method, as the reason for their withdrawal.

These results concur with prior research that emphasizes the importance of a supportive learning environment to the success of many women students (e.g., AWM, 1989; Davis, et al., 1996). Also further supported by this work is evidence from a number of studies indicating that minority students excel in environments that foster the development of peer support networks such as those developed in cooperative learning settings (e.g., Ganter, 1994; Treisman, 1985; Valverde, 1984). There is evidence from one current study that minority students also share some of the same concerns about reform calculus as their majority peers, including the value of the extra time involved in learning about technology and the thought that more traditional courses are "real learning." (Eiseman, et al., 1996) More research is therefore needed to clarify the various ways in which calculus reform affects women, minorities, and other specific populations.

Studies on student attitudes as impacted by calculus reform yielded several other results that span institutional types and reform methods. Areas of particular interest include changes in student confidence, causes of student resistance to reform, reactions to reform textbooks, the impact of the classroom environment, and student perceptions of the value and relevancy of reform. Trends observed in each of these areas are discussed below.

B.3 *Changes in student confidence levels*

One of the most profound areas of impact as a result of reform calculus is the effect on student beliefs about their ability to succeed in mathematics. Significant positive increases in attitudes about their capacity to think mathematically and to contribute valuable insights to problem-solving endeavors were observed at a number of institutions (Armstrong, Garner, and Wynn, 1994; Bonsangue, 1994; Bookman & Friedman, 1998; Crocker, 1991; Ganter & Jiroutek, 2000; Morstad, 1996; Pustejovsky, 1996; Salem, et al., 1996). This is not to say that the students do not find reform courses (as well as traditional courses) difficult, challenging, and often overwhelming (Cribbs, et al., 1996; Eiseman, et al., 1996; Newell, 1994). It simply means that, given the support they need, students are able to develop mathematically in ways that do not typically occur in many traditional classrooms. This personalization of mathematics also enables students to display more confidence and satisfaction in their abilities to do calculus (Frid, 1994). This increase in confidence levels naturally leads to a more positive attitude about mathematics, an understanding of its importance, and a greater likelihood that these students will enroll in more mathematics courses after calculus (Allen, 1995; Armstrong, Garner, and Wynn, 1994; Cribbs, et al., 1996; McCallum, 2000; Morstad, 1996; OuYang & Hann, 1996). However, it is important to note that these reactions are not universal (Eiseman, et al., 1996; Hershberger & Plantholt, 1994; Pustejovsky, 1996).

B.4 *Causes of student resistance to reform*

As previously discussed, certain student audiences have reacted in various ways to the reform efforts in calculus. A number of studies further investigated the reasons for negative student reactions. In particular, several components of reform contributed to the resistance of students in these courses. For example, some complaints focused on technology, including the belief that it was unfair to burden students with the additional workload of learning to use computer software packages or a graphing calculator in addition to the calculus content (Eiseman, et al., 1996; Monteferrante, 1993; Salem, et al., 1996; Silverberg, 1994). Similar complaints were voiced for reform courses that implemented group work or long-term projects, with students stating that they need more "pencil-and-paper stuff" and other "real mathematics." (Eiseman, et al., 1996; Ganter & Jiroutek, 2000; Monteferrante, 1993; Pustejovsky, 1996; Salem, et al., 1996) Students were also concerned that the course was designed with the intent of accomplishing too many things; e.g., the use of technology (as well as a focus on technical writing) would necessarily imply that their computational skills would suffer (Monteferrante, 1993; Newell, 1994). These complaints can often be successfully addressed with continued experience and open communication about the goals of the course with the students (Hershberger & Plantholt, 1994; Newell, 1994; Pustejovsky, 1996).

Availability of the required technology is also a source of trouble for students in reform courses. The convenience of graphing calculators that can be transported anywhere and that are not likely to suffer from "crashes" often makes them more appealing to students than computers (Prettyman, 1996). This preference is in spite of the fact that calculators add to their out-of-pocket expenses, while the more powerful computer technology is often available for free and within walking distance of their residence. This attitude is changing, however, as computers become more reliable and even more readily available to students (Prettyman, 1996; Salem, et al., 1996).

Group work also results in very mixed reactions from students, especially when combined with assignments on which they receive a group grade. Again, the traditionally strong mathematics students are often the ones who resist this type of work, as they feel it is unfair to be "burdened" with the less-prepared students and perceive that working with them creates the risk of receiving a lower grade on the assignment (Cribbs, et al., 1996; Monteferrante, 1993). The value of communicating mathematics and discussing problem-solving strategies with their peers is not understood by many of these students (Hershberger & Plantholt, 1994). This attitude can change (and usually does) by the second term of the course, when students begin to see how much they have grown mathematically. Believing that the ideas they communicate with their peers are valid is a critical part of positive group experiences for students (Packel, 1996; Pustejovsky, 1996).

B.5 *Student reactions to reform textbooks*

Textbooks have long been the major driving force in most calculus courses. Therefore, leaders in the calculus reform efforts have placed a heavy emphasis on the development and publication of texts that would help in the dissemination and promotion of their work. The result has been a wide variety of textbooks that support various aspects of reform, ranging from fairly conservative to very different changes in the curriculum. The feedback presented here represents textbooks that are somewhere in between, with a tendency toward the conservative end of the spectrum.

The most common statement from students is that the reform textbooks are not easy to read (OuYang & Hann, 1996; Pinter-Lucke, 1996). Students also readily admit that they rarely have read the textbook in previous mathematics courses and that their problems with reading the book were actually the result of not knowing how to read mathematics. Related negative comments, such as "the book does not have enough examples for me to do the homework" and "there are no answers in the back of the book," also may be the result of students that have not been exposed to mathematics in a way that the reform texts require (Lepowsky, 1996; Pustejovsky, 1996). It is important to note that authors for many of the reform textbooks have responded to these concerns in subsequent editions by including more examples and solutions for select homework problems. It is also interesting that many more traditional calculus textbooks have been revised to include more applications and problems that utilize technology, implying a (partial) "meeting of the minds" on the desired content of a calculus textbook between authors of reform and traditional textbooks.

Therefore, although the feedback from students about reform textbooks is valid and should be addressed, it is very important not to let these comments result in a comparison war between reform texts and traditional ones. Rather, mathematicians need to make decisions about what the textbook in a calculus course should help students to accomplish and work to develop texts that achieve these goals. Helping students learn to read mathematics effectively and to use the information presented in a calculus book to further their understanding of the subject needs to be an integral part of any calculus course.

B.6 *Impact of classroom environment*

The general learning environment created in a reform course can have a significant effect (positive or negative) on the attitude of students. In general, because students perceive that reform courses involve more work than traditional courses and are harder because of the frequent use of open-ended problems and reading assignments, they express a strong need for organized help in the form of additional faculty office hours, more question and answer sessions with teaching assistants, organized study groups, and extra time in the computer lab (Bookman & Friedman, 1998; Newell, 1994; Pustejovsky, 1996; Salem, et al., 1996). This need decreases as students become more experienced with the reform environment (Crocker, 1991). The integration of classroom and laboratory experiences is also very important. When this connection is not made, students often become very frustrated because they do not see the relevance of labs that are seemingly unrelated to the other course material.

The experience and comfort level of the instructor with the reform method is important to the students' perceived success of the course. When both the instructor and the students are novices with the method, it creates an atmosphere of uncertainty and confusion. However, this is a temporary condition, with the attitudes of both students and faculty improving after one or two semesters of experience with the ideas of reform (Bookman & Friedman, 1998; Hershberger & Plantholt, 1994; Prettyman, 1996; Pustejovsky, 1996; Salem, et al., 1996; Silverberg, 1993; Wang, 1996). This improvement in attitudes is especially pronounced when the instructor has revised the course based on continuous student input. Therefore, *the instructor and the instructional methods—not the particular reform curriculum or textbook adopted—are the real crux of reform.* An important element of this is the perception of students that they receive more personalized attention in reform courses.

B.7 *Student perceptions of the value and relevancy of reform*

A number of studies have reported student perceptions on the value and relevancy of reform. Many students have expressed that they feel more positive about the active role that they are able to take in the learning process while in reform courses (Armstrong, Garner, & Wynn, 1994; Bookman, 2000; Jackson, 1996; Keith,

1995; Prettyman, 1996). Students also feel that the more visual approach taken by some reform courses is very helpful. The technology, conceptual problems, and applications that support this visual approach are also important and help to make calculus more interesting and understandable (Andrews, 1997; Bookman & Friedman, 1998; Burgess, 1996; Cribbs, et al., 1996; Ganter & Jiroutek, 2000; Keith, 1995; McCallum, 2000; Packel, 1996; Prettyman, 1996; Salem, et al.; 1996; Soto-Johnson, 1996).

Students claim that they study with greater enthusiasm, to the point that they often work independently on unassigned problems and voluntarily obtain graphing calculators so they can explore problems on their own (Armstrong, Garner, & Wynn, 1994; Morstad, 1996; Walker, 1996). Cooperative computer labs and the responsibility to others that is implied by group work encourage students to participate while in class, as well as spend more time studying calculus outside of class (Bonsangue, 1994; Pinter-Lucke, 1996; Prettyman, 1996; Soto-Johnson, 1996). In general, students find reform courses more challenging, but worthwhile (Ganter & Jiroutek, 2000; Pustejovsky, 1996; Salem, et al., 1996). They often perceive that they are learning more and will be better prepared for future coursework and jobs (Armstrong, Garner, & Wynn, 1994; Morstad, 1996; Narasimhan, 1993; OuYang & Hann, 1996). They also are quick to point out that the quality of pedagogy in a reform course, including the effectiveness of the instructor, is at least as important as what is presented (Bonsangue, 1994; Eiseman, et al., 1996).

C. Conclusions

As with student achievement, there have been many changes in student attitudes as a result of calculus reform. In general, the confidence of the reform students in their conceptual understanding, problem solving skills, ability to visualize, and integration of topics is significantly higher than that of their traditional peers. Reports from some institutions indicate that reform students are enrolling in more non-required mathematics courses beyond calculus than their traditional peers, implying that perhaps calculus reform generates more interest in mathematics.

However, students' attitudes about the role of more traditional skills in the calculus course have not changed much as a result of reform. For example, while students feel that the use of technology is important, they also think that pencil-and-paper computations must continue to be emphasized. In fact, one of the concerns expressed by students about the reform course is that their computational skills are no longer as strong as they should be.

These and other concerns raised here are critical to the future of reform and should be further investigated to determine their impact on calculus students. Specifically, the following questions are raised as important topics for future study:

- Two independent studies from institutions using the same reform curriculum in all sections of the course, studying students with similar high school experiences, reported conflicting results regarding the difficulty that these students had in adopting to reform at the college level. This result emphasizes that, although their background and current environment are the same, two students may still walk away with different perceptions of a calculus course. Are these perceptions real or imagined?

- What are the various ways in which calculus reform affects the attitudes of women, minorities, and other specific populations?

- Student feedback on reform calculus textbooks was generally negative. Why? And although the feedback from students about reform textbooks is valid and should be addressed, is the comparison of reform and traditional texts really the appropriate comparison? Or should we make decisions about what the textbook in a calculus course should accomplish and work to develop texts that achieve these goals? If so, are such decisions possible? In addition, helping students learn to effectively read mathematics and to use the information presented in a calculus book to further their understanding of the subject needs to be an integral part of any calculus course. How can this best be accomplished?

- The available data seem to indicate that the instructor and the instructional methods—not the particular reform curriculum or textbook adopted—are the real crux of reform. Is this true? If so, what role does the perception of students that they receive more personalized attention in reform courses play in this result?

Faculty Reactions

A. The nature of available information about faculty

Although some institutions implementing various reform methodologies chose to investigate the impact of those changes on faculty, very few formal studies exist. The nature and degree of faculty reactions, including the impact on their teaching as well as their attitude toward alternative teaching styles, is a critical component of reform and will ultimately determine the long term sustainability of these efforts. Therefore, in spite of the lack of "concrete data" on the subject, this chapter is devoted to a discussion about faculty reactions to calculus reform.

As part of the data collection for this report, a letter was sent to approximately 600 individuals who have contributed to the reform efforts (see Appendix C). The letter requested assistance in the compilation of existing evaluation studies in calculus reform (see Chapter 2, Section C). Although the letter did not specify information on any particular component of the reform efforts, many individuals (especially mathematics department chairs) chose to discuss at length the changes in the departmental environment as a result of calculus reform. Specifically, the letters from these individuals mentioned dramatic changes in faculty attitudes and interests as related to teaching undergraduate mathematics courses. The amazing fact about this information is that, without guidance, these individuals made virtually the same observations about the changes within their respective departments. These submissions were made by individuals from almost 50 institutions, representing the entire spectrum of institutional types, geographical locations, and reform methodology.

For reasons of confidentiality, the authors and institutions from which these letters were submitted cannot be published here. However, the following quotes provide a sampling of the similarities observed in these submissions:

The impact of reform

Most important to us is the iterative process of reviewing and revising our calculus curriculum that has gone on continuously since we adopted reform materials.

[T]here has been a tremendous increase in departmental concern about, and involvement in, the teaching of calculus since we adopted the reformed materials.

[Our project] made a significant impact on the department and on the university.

Standards for calculus

[I]t is necessary to develop a national standard as a guide for teaching calculus.

I suggest that NSF appoint a curriculum committee to develop a standard [calculus] curriculum that can be recommended to or adopted by universities and colleges.

Reactions of faculty

Not all of our colleagues were as convinced as we were of the value of the computer, appropriately used, in the teaching of calculus.

Attitudes of our faculty toward the reform movement seem to range through the full spectrum.

About a fifth of our faculty seem to be solidly against the reform movement with varying degrees of passion.

[Faculty] have attitudes ranging from indifferent to nervous to curious.

The need for evaluation

I have countless times been asked if I could cite statistics or evidence that calculus reform is really working.

I hope that I will see an evaluation of these efforts in the near future.

A common test for students having had a traditional course and students having had a reform course clearly will not do since the objectives are different and any such common test would necessarily be unfair to one group or the other.

The letter of request sent to the community as part of this project was not intended to provide a forum for individuals to submit their opinions about the impact of reform. As such, the original plan for this report did not include any provisions for this type of subjective information. However, the number of letters with remarkably similar statements about faculty reactions to reform was so overwhelming that to ignore this information would have diminished its importance as an outcome of the reform efforts. Therefore, this chapter conveys the main points made in these letters in combination with the results from the existing studies on the impact of reform efforts on faculty.

The challenges of reporting soft data such as that from these individual letters are numerous and inevitably affect the reliability with which the results can be reported. The conclusions made here are the interpretation of this author and therefore are subject to question. The information is certainly not meant to be definitive or complete in its presentation here. However, it is important to consider this information, even if just within the scope of its potential meaning to individual institutions.

B. The impact of current reform efforts

For the purpose of this study, faculty reactions are defined to be results from studies designed to measure activities and attitudes of individual faculty and/or entire departments about efforts in calculus reform. These results include reactions both from faculty with experience teaching calculus using reform ideas and those without. Methods used by the investigators to collect this information include faculty surveys and interviews, external review teams and other peer observations, conference and workshop evaluations, faculty journals, and logged discussions at department meetings. The analysis of NSF files revealed that 27 of the 127 funded projects reported results from the collection of such data on faculty attitudes. Numerous other studies at NSF-funded sites are being conducted, although the information is not yet available. Forty-two additional studies on faculty attitudes that were not funded by NSF were also reviewed (see Table 14 for a summary of the studies analyzed in this chapter). Included in these additional studies are the relevant findings from an external evaluation of the NSF Course and Curriculum Development program, the broader program from which the calculus projects were funded (see Eiseman, et al., 1996).

The variations in faculty responses to reform are enormous and it is therefore very difficult to extract definitive trends from the information. A few common areas addressed by a number of studies will be discussed here. These include changes in

- departmental and college-wide environments
- mathematics courses before and after calculus, and
- individual support for reform ideas.

Faculty reactions to technology, cooperative learning, and other alternative classroom environments will be discussed.

B.1 *Changes in the mathematics environment*

The most important statement to be made in terms of the environment in mathematics departments as a result of the reform efforts is that there has in fact been change. Although this may seem an obvious outcome, it is in reality a major breakthrough that is clearly observable in many departments at a wide variety of institutions. Of course, not every change has been viewed by faculty as positive; nonetheless, for better or worse, change in the environment has occurred. Specific examples of such change that have been widely documented include

- increased conversations within departments about undergraduate mathematics education,
- increased need for (and often actual) conversations with faculty in other disciplines to coordinate the curriculum in mathematics with other subjects,
- the "domino effect" that change in calculus has had on the teaching of other undergraduate mathematics courses, and
- increased interest and participation in educational issues by mathematics faculty.

Conversations and collaborations among mathematics faculty and with faculty in other departments have been encouraged by reform. These conversations often result from the changes implied by the new methods for the types of problems that are the focus of the newly-developed curricular materials (Keynes, Olson, O'Loughlin, & Shaw, 2000; Lepowsky, 1996; Mumford, 1997). For example, the increase in applications has made necessary a closer tie between calculus and the related topics in physics, chemistry, biology, and a host of other disciplines. The changes in calculus have also made it difficult to ignore the structure of other mathematics courses, especially those that are often subsequent to calculus, such as differential equations and linear algebra (Collins, 1996). Efforts in mathematics reform have logically moved into these courses, as well as precalculus, with funding from NSF supporting this migration. Naturally, this means that faculty who have traditionally taught these courses within their departments are now discussing topics and issues

Table 14. *Sources of evaluation data for faculty reactions by Carnegie classification* and NSF funding.***

Source***	Research/doctoral		Comprehensive		Liberal Arts		Two-Year		High School		Total	
	NSF	Non-NSF	NSF	Non-NSF	NSF	Non-NSF	NSF	Non-NSF	NSF	Non-NSF	NSF	Non-NSF
Published article	4	11	0	6	1	3	0	2	0	0	5	22
Published report/ collections of articles	0	4	0	1	0	0	0	0	0	0	0	5
Conference proceedings	1	1	1	0	0	0	0	0	0	0	2	1
Unpublished manuscript	0	0	1	0	2	0	1	0	0	0	4	0
Doctoral dissertation	4	2	0	0	0	0	0	0	0	0	4	2
Masters thesis	0	0	0	0	0	0	0	0	0	0	0	0
Internal institutional report or report to NSF	2	0	0	0	1	0	1	0	1	0	5	0
Individual letter	1	6	3	1	2	2	1	0	0	3	7	12
Total	**12**	**24**	**5**	**8**	**6**	**5**	**3**	**2**	**1**	**3**	**27**	**42**

The header spanning all frequency columns reads: **Frequency (*n* = 69)**

* Indicates the Carnegie classification for the primary institution where the work was conducted.

** Indicates whether the document was produced as part of a NSF-funded project.

*** For complete citations on all references used in the compilation of data for this project, see the MAA website (www.maa.org).

that reach across course boundaries. Consequently, the structure of the entire undergraduate mathematics curriculum, including requirements for the mathematics major, are being reviewed in numerous departments (Buccino, 2000; Dossey, 1998; Keynes, et al., 2000; Jackson, 1996).

The activities of many mathematics faculty, both within their departments and in the larger professional community, have changed dramatically over the past decade. An indisputable indicator of this change is the large (and increasing) attendance and participation in special educational sessions at the national and regional meetings of the American Mathematical Society, the Mathematical Association of America, and the Society for Industrial and Applied Mathematics. Virtually non-existent at meetings in the mid-1980s, these sessions now comprise a large part of conference programs and are often filled to capacity. Attendance also has increased at conferences that specifically focus on change in undergraduate mathematics education, such as the International Conference on Technology in Collegiate Mathematics and the Annual Conference on Research in Undergraduate Mathematics Education. Within departments, faculty now struggle with the traditional structures that often do not encourage and support such activity in mathematics education. Guidelines for tenure and promotion are being questioned and many involved in reform, especially untenured faculty, are anxious to learn how the mathematics community will respond to the changing professional roles that are implied within departments committed to curricular innovation.

B.2 *Faculty reactions to alternative teaching environments*

Faculty who enjoy financial and collegial support report a positive impact of reform methodologies, especially technology, on their teaching and professional development (Collins, 1996; Eiseman, et al., 1996; Keynes, et al., 2000; Waggoner, 1996). On the other hand, a lack of support from administration and fellow faculty members, whether technical, moral, or financial, often leads to demise of new methods (Burgess, 1996; Cipra, 1996; Eiseman, et al., 1996; Pence, 1994). The aspects of reform that faculty find most valuable for students, as well as most rewarding—and essential—in their teaching are

- use of technology
- teaching in a more collaborative environment (both with their students and other faculty), and
- opportunities for professional development that have come with reform (Cribbs, et al., 1996; Eiseman, et al., 1996; Keynes, et al., 2000; Mathews, 1995; Prettyman, 1996; Pustejovsky, 1996).

In general, faculty believe that some combination of these methodologies helps students to develop a strong conceptual understanding of the material while also nurturing positive faculty attitudes about their students' ability to do mathematics. Of course, these positive aspects of reform methodologies are also the ones that produce some of the greatest difficulties, including

- how to manage a collaborative learning environment, and
- how to change the primary mode of assessment from student testing to some combination of this and alternative forms of assessment that will truly measure the learning of individual students (Andrews, 1996, 1997; Buccino, 2000; Keynes, et al., 2000; Lepowsky, 1996; Prettyman, 1996).

The dramatic changes in the nature of calculus courses—and other mathematics courses—that utilize the ideas of reform have also created a level of discomfort among some faculty that has led to questions of academic freedom and how to handle dissenting colleagues. Such questions include the following:

- If a department is in general agreement about the content and structure of courses (reform or not), should faculty that are not in the majority be forced to change their style? If so, how should this be done?
- Is it important that all faculty within a mathematics department teach freshman calculus, or is it better to simply let the "converted" teach all these courses? If so, what happens when these faculty want to teach other courses? Will other faculty then be brought in to cover the calculus course, implying that the course is no longer taught in the reform spirit?

- In reference to the above questions, how do you deal with issues related to student equity if a course with large campuswide enrollments such as calculus is taught in dramatically different ways by the different instructors?

- What role should graduate students and temporary/part-time faculty have in the teaching of calculus? How can departments give these individuals incentive to participate in curricular changes and ensure that they are implementing the spirit of the new methods?

- What responsibility does a department have in providing faculty development opportunities that help instructors adapt to new classroom environments? Who should bear the cost of such training, especially when it involves large numbers of temporary faculty and graduate students who are constantly changing?

- Should faculty working on curricular reform receive explicit departmental support for the large commitment of time and energy necessary to affect such changes? If so, what form should that support take? (Armstrong, Garner, and Wynn, 1994; Burgess, 1996; Cribbs, et al., 1996; Eiseman, et al., 1996; Jackson, 1996; Keynes, et al., 2000; Lepowsky, 1996; Pence, 1994; Pustejovsky, 1996)

Some faculty are also concerned that the reform methods imply the elimination of certain fundamental mathematics skills that will hinder students in future courses. Specifically, the impact of computers and graphing calculators on students' ability to calculate with pencil and paper is a concern for many faculty (Andrews, 1996, 1997; Cipra, 1996; Prettyman, 1996; Wang, 1996), as is the use of technology simply as a means for performing technical calculations (Eiseman, et al., 1996). These concerns naturally motivate discussions about the relative importance of computational skills completed by hand as compared with the alternative skills nurtured by technology.

C. Evaluations: external review teams and faculty development workshops

For the purpose of this study, an external review team is defined to be a group of professional colleagues (usually three in number) that are asked to visit an institution and observe the calculus reform program in action. The task assigned to these teams is to analyze the methods being implemented in the classroom and determine the impact of the changes on students, faculty, and the general learning environment. Each team employs some combination of classroom observations, review of course materials, and conversations with administrators, faculty, and students to complete this task. The team then discusses their experiences among themselves and submits a report of their findings to the project directors at the institution observed. This information is then shared with NSF, usually as part of the final report for the project. The analysis of NSF files revealed that 18 of the 127 funded projects submitted such data from external review teams (see Table 8). Results most frequently reported from the external review teams include the following:

- faculty are re-energized

- computers involve more preparation on the part of both faculty and students, but the changes implied in the learning environment are very positive

- students need significant help in synthesizing ideas developed when learning in alternative environments

- when interviewed, students give mixed reviews on technology, citing that they believe they are learning more but resent the extra time and energy involved

- appropriate institutional infrastructure is critical to success, including administrative support, clear communication between departments, and well-defined requirements for transferring between courses

- students are learning some very different skills, such as technical writing and working in teams; grading standards and student workloads must be adjusted to accommodate these changes

- the nature of the reform textbooks implies a greater need for students to read mathematics; departments need to address this issue in ways that encourage and support the reading of mathematics by students.

Workshop evaluations include reactions from participants in workshops and conferences focusing on the implementation of calculus reform ideas. These reactions were collected by the host institution, usually on a survey distributed at the end of the workshop or conference. This information was then shared with NSF, typically as part of the final report for the project. The analysis of NSF files revealed that 11 of the 127 funded projects submitted such evaluation data (see Table 8). Results indicate that, in general, faculty find these professional development opportunities very helpful in the implementation of new ideas in their own classrooms. Specifically, working directly with their colleagues who have developed and/or have experience with the methods—while also interacting with others who are just beginning to make changes—helps them avoid the mistakes made by others and build a network of colleagues that they can continue to work and talk with after returning to their home campuses. Follow-up interviews conducted with a subset of the participants for one workshop (25 of 800) revealed that 52% had changed how they used technology and 62% had changed pedagogical practices in their calculus courses. The most common dissatisfaction expressed by participants is that they need even more professional development opportunities, especially in the use of technology. These results imply that small workshops and conferences, especially those led by very experienced colleagues, are perhaps the most effective means of disseminating ideas and encouraging change across campuses.

D. The "backlash" against reform

A discussion focused on faculty reactions to calculus reform would not be complete without an acknowledgment of the dissent expressed by a significant group of faculty nationwide. This dissent, commonly known as the "backlash" against calculus reform, is based on the belief that the major components of the reform movement (technology, alternative pedagogical techniques, and an emphasis on applications) seriously undermine the primary goal of the calculus course: to teach students mathematics. Reform courses have been labeled by some as "fluff," "soft," and "watered-down mathematics."

However, many mathematicians opposed to the reform efforts are not opposed to change per se. They agree that the calculus course needs to be reconsidered. The aspect of reform on which they disagree is the way in which the current efforts have addressed this need. Some say that the real problem with calculus (and undergraduate mathematics in general) is not the course itself. Specifically, many mathematicians believe that the major problems to be addressed are

- **the attitudes of students:** specifically, students have been led to believe that learning should not be hard work—and undergraduate mathematics courses reinforce that belief by de-emphasizing homework, independent study, and a commitment to learning (Andrews, 1995; Cipra, 1996; Culotta, 1992; Johnson, 1997; Klein & Rosen, 1997; Norfolk, 1997);

- **the size of calculus classes:** there is considerable evidence that personal contact with an instructor can greatly enhance a student's educational experience, as well as their performance in the course; calculus is a difficult course for many students and therefore a nurturing and supportive environment is critical to success (Andrews, 1995; Keynes, 1991; Zorn, 1991);

- **the appropriate course goals and means for evaluating success:** the need for evidence that reform methods help students to better understand mathematics has been stated repeatedly; those opposed to reform ask the question "Why?" and even those involved in reform projects want desperately to know if their efforts are having a positive impact on students (Andrews, 1995; Buccino, 2000; Culotta, 1992; Keynes, 1991; Klein & Rosen, 1997; Zorn, 1991).

In addition, a major impediment to change in the calculus course is the de-emphasis on the importance of teaching at many colleges and universities. With many activities competing for faculty time, the rewards associated with those activities have a great impact on the choices faculty make (Boyer, 1987, 1990; JPBM, 1994; Keynes, et al., 2000; NRC, 1996). Less obvious, but just as important, in affecting the decisions that faculty make is the local culture beyond the reward structure; that is, is innovative teaching encouraged and supported? Specifically, does the institution provide opportunities for professional development that will foster excellence in teaching? Without appropriate training, faculty cannot implement the wealth of existing ideas about effective teaching—whether those ideas be "reform" or "traditional" (Keynes, et al., 2000; NRC, 1996; see also Chapter 6, Section B.2).

Finally—and perhaps an even greater concern than those outlined above—is the line that has been drawn between "reformers" and "traditionalists." One of the most positive results of the reform efforts has been a renewed discussion about undergraduate mathematics (see Chapter 6, Section B.1). Even debates about the changes proposed are healthy when based on fact and common goals. Unfortunately, many such debates have deteriorated to slander and personal attacks from both "sides." (*Chronicle*, 1997; Klein & Rosen, 1997; Wilson, 1997) These arguments have done nothing to promote the positive work of the mathematics community in undergraduate education and have led some individuals in other arenas to think that mathematicians do not know what they are doing, ending in a plea that they just "please fix it quickly—we need the calculus." (Zorn, 1991, p. 1) Perhaps the lesson to be learned from the heated discussions about calculus reform is that "no single calculus course can—or should—be all things to all people." (Zorn, 1991, p. 4) Many agree that a uniform set of broad goals for the course is needed. The implementation of those goals could—and should—be achieved in many different ways to meet the local needs of faculty and students (Reynolds, 1991; Zorn, 1991).

E. Conclusions

Faculty are perhaps the most instrumental group in the development and implementation of curricular change. Ultimately, the level of faculty support for such innovations will determine the long-term success or failure of educational reform at any level and in any course. Therefore, it is important to consider their needs and concerns when making such changes.

The observation on which there is no disagreement is that there has in fact been change. Whether that change has been positive or negative is the source of heated debate. However, most faculty agree that some kind of change in the calculus course is necessary. An additional idea on which there is no dissent is the primary goal of the calculus course: to help students have a better understanding of and appreciation for mathematics. Of course, individual definitions of "understanding" and "appreciation" vary dramatically. Finally, most agree that an environment of change—however that is defined—is here to stay. Calculus reform has led to changes not only for calculus, but for the entire undergraduate mathematics curriculum. And, these changes have meant increasing discussions with colleagues in other disciplines, as well as secondary mathematics teachers. The continually increasing attendance at educational sessions at mathematics conferences attests to the fact that mathematicians think these discussions are important—even if they do not agree on the outcomes.

The research on faculty reactions to reform is very limited and many questions still need to be answered (see Chapter 6, Section B.2). Faculty continually lament the need for more and better opportunities to learn about alternative teaching strategies, as well as methods for appropriately assessing student learning in these environments. Colleges, universities, and professional societies must assume the responsibility for providing these professional development opportunities if innovation and excellence in undergraduate mathematics education is to be sustained in the next decade.

Discussion and Conclusions

A. Current status of reform

Almost 200 evaluation documents, plus information from NSF files on 127 funded projects, were compiled to produce the information in this report. These documents include published articles, doctoral dissertations, conference presentations and proceedings, letters describing results from projects, internal reports submitted within various colleges and universities, and a host of other materials. All of this information was summarized in a qualitative database and synthesized, resulting in a wealth of information on the decade of calculus reform between 1988 and 1998. An overall review of the information submitted reveals the following trends.

1. Pedagogical techniques most often cited as part of the projects are
   - technical writing
   - discovery learning
   - use of multiple representations of one concept, and
   - cooperative learning.

2. The most popular methods for distributing results to the community are
   - conference presentations
   - journal articles
   - organization of workshops, and
   - invited presentations at college colloquia.

3. Evaluations conducted as part of the curriculum development projects revealed
   - better conceptual understanding, with the effect of reform on computational skills uncertain
   - higher retention rates
   - higher scores on common final exams, and
   - higher confidence levels and involvement in mathematics for students in reform courses versus those in traditional courses.

4. In general, regardless of the reform method used, the attitudes of students and faculty
   - seem to be negative in the first year of implementation, and
   - steadily improve in subsequent years if continuous revisions are made based on feedback.

5. There does not seem to be any consistency in the characteristics of faculty who are for/against reform; however,
   - students who seem to respond most positively to reform have little or no experience in calculus prior to entering the course, and
   - students who seem to respond most negatively to reform are traditionally at the top of their high school class.

6. The reform effort has motivated many (often controversial) conversations among faculty about the way in which calculus is taught; these conversations are widespread and continuous and they have resulted in a renewed sense of importance about undergraduate mathematics education.

7. Most faculty believe that the "old" way of teaching calculus was ineffective; however, there is much debate about whether we are moving in the right direction.

8. Changes in the requirements of standardized tests seem to have enormous effect on the practices of secondary Advanced Placement (AP) calculus teachers; that is, the requirement of graphing calculators on the AP calculus exam since 1995 has ensured their use in virtually every secondary calculus classroom, even though many AP teachers are opposed to such a requirement.

9. Faculty universally agree upon the importance of evaluating the effect of the reform efforts on
   • student learning
   • faculty and student attitudes, and
   • curriculum development.

   This information is needed to
   • justify their work within their department
   • understand the impact of various methods, and
   • give them motivation to continue in the struggle.

Calculus reform efforts have yielded mixed results in the areas of student achievement and student attitudes. The varying outcomes at different institutions are certainly not surprising, given the wide range of definitions of "reform" that have been implemented by each of the projects. Nonetheless, a number of elements are in fact common to many of the projects, including the use of computer technology and applications in the teaching of calculus. Many participants in the reform efforts also believe in the importance of emphasizing a student-centered environment, including discovery learning and cooperative activities that support a more conceptually-based course.

The existence of these common elements throughout the majority of the projects implies that *the relative success or failure of reform efforts is not necessarily dependent upon what is implemented, but rather how, by whom, and in what setting.* The consistent reactions of students from a wide variety of institutions point to several key components in the success of a reform environment. For example, instructors must communicate to students (and other faculty) the purpose of the changes being made in the calculus course. This is perhaps not as easy as it seems, as the reasons for the change must be seen as relevant and important to future success. It is unfortunate that many students believe mathematics is a static list of rules and algorithms to be memorized, a barrier to be overcome before they can do "real" problems in other disciplines. Perhaps the most important role of reform efforts then is to challenge these beliefs and help students to see the many uses of calculus, both within the discipline of mathematics and in a wide range of problems from other disciplines. However, the means by which this can be communicated is not at all clear and needs to be addressed by the mathematics community as a critical part of the reform efforts.

The level of personal attention available to students also greatly affects their attitude and level of commitment in a calculus course. This is not a characteristic unique to reform courses, but one which has been highlighted as a result of the foreign environment that a reform course introduces. Specifically, the elements that define such a course are often ones that students have never experienced in mathematics (or perhaps any other discipline) and therefore additional support is required as the students adjust their learning styles. This is likely to be the reason that many of the projects report a dramatic shift in student (and faculty) attitudes after they have experienced a couple of terms with reform calculus.

One area of particular concern is the adamant opposition to reform of many students who have excelled in the traditional environment. It is important that the reform efforts not cause these students to lose their interest in mathematics, just as it is important that others be encouraged through the wider variety of opportunities for success that reform courses offer. It is not necessarily the case that these two goals must be in opposition. As with anyone who is opposed to change (including faculty), it is likely that these students are simply reluctant to move from a learning environment in which they are apparently successful to one which is unfamiliar. Although this can be overcome with open communication and creative opportunities that continue to challenge these students, it is certainly not something that can be deemed as unimportant because "these students will succeed no matter what we do with them." A primary goal of the reform efforts is the creation of a course that makes calculus real and interesting for *all* students, including the ones for whom it has always been, as well as those for whom it has not.

B. Areas for further study

The following are suggested as areas for immediate further study (see also Ganter, 2000).

- What exactly have we learned—about calculus, about undergraduate mathematics education, and about the interaction of the discipline of mathematics with other courses in the sciences and beyond?

- How can this knowledge be applied in positive ways to the organization and delivery of undergraduate mathematics courses while maintaining the mathematical integrity of these courses?

- What are the contributions that mathematics can—and should—make in the world of science, engineering, and technology? How can we better nurture our relationships with colleagues in these other disciplines?

- What mathematical skills and knowledge should an undergraduate student have after completing a first-year collegiate mathematics course?

- What has been learned about the cognitive processes involved in learning mathematics through the extensive research conducted in this area during the past decade? How can we use this research to better inform the development of undergraduate mathematics courses?

- How does mathematics fit into the broader context of student learning; e.g., what are the life skills that our students will need to succeed in the workplace?

- What is the appropriate role of technology in the teaching and learning of mathematics?

- How does a first-year collegiate mathematics course contribute to the overall mathematics education of our students? How can the changes made in this course be used to improve other mathematics courses?

- What is the appropriate role of colleges and universities in supporting curricular change? How can administrators provide an environment that is conducive to change and enables faculty to develop the necessary skills for supporting such change?

- What are the appropriate mechanisms by which the mathematics community can evaluate progress and thereby better inform continuing change? What changes ultimately help our students to better understand mathematics and to have an appreciation for the importance of mathematics in our society?

And, finally

- What questions still need to be asked and answered, through carefully conducted research, to guide future changes?

A major purpose of the calculus reform movement has been to revitalize calculus and generate discussions within the mathematics community about the nature and content of the calculus course. Indeed, this has happened and is a major accomplishment that should not be overlooked. A question that naturally comes out of such discussions is whether the reform efforts are in fact helping students to better understand calculus and appreciate the importance of mathematics in their lives. This is a challenging question that may never be definitively answered, but one which must be studied if calculus courses—and mathematics in general—are to be a vital part of the undergraduate curriculum. This project represents one effort to evaluate the national impact of the calculus reform movement, the results of which provide a baseline not only for the evaluation of reform in undergraduate mathematics, but also for future efforts in calculus. And although this study adds to the knowledge base on the impact of calculus reform, it is only the beginning. More questions have been generated by this work than answered and the mathematics community must therefore continue such investigations as part of the efforts to renew calculus and the undergraduate mathematics curriculum.

Appendices

A. Framework for calculus reform database
B. Indicators of evaluation quality
C. Letter of request to the mathematics community
D. Definitions for national objectives of calculus reform

Framework for Calculus Reform Database

I. General information

- Name of institution and general statistics (size, location, Carnegie classification, etc.)
- Number of students taking reform, with breakdown by ethnic/gender types
- NSF funded? Y/N   Other funding? Y/N   By whom? _____
- Amount of funding (if known)
- Other _____

II. Abstract

III. Student audiences targeted by reform efforts (circle all that apply)

Postsecondary students	Females
Secondary students	Minorities
Non-math majors	Other _____

IV. National objectives of reform[1] used by project (circle all that apply)

Technical writing	Use of graphing calculators	Use of computers
Conceptual understanding	Modeling (construct/analyze)	Applications
Cooperative learning	Open-ended problems	Approximation
Oral presentations	Mathematical reading	Discovery learning
Extended-time projects	Problem-solving skills	Laboratory experience
Alternative assessment	Focus on differential equations	Other _____

Multiple presentations of one concept (symbolic/verbal/numerical/graphical)

V. Changes in educational environment[2]

Discussions among institution's faculty: within mathematics and with other disciplines
Institutional commitment of resources, etc.
Other courses changing as a result of the calculus project
Other _____

---

[1] Developed in part from Tucker (1990), Tucker & Leitzel (1995), Roberts (1996), and Schoenfeld (1996).

[2] This section does not apply to analysis of NSF files reported in Chapter 3.

VI. Evaluation activities (circle all that apply in each column and include any results)

Planned activity	Actual	Results
Retention/completion rates	Y/N	
Achievement/GPA		
Experimental vs. control (exams, etc.)		
Student survey/attitudes		
Student interviews		
Faculty survey/attitudes		
Faculty interviews		
Site visit/external review team		
Laboratory evaluations		
Course evaluations		
Longitudinal study		
Workshop evaluations		
Cost effectiveness		
Disadvantaged student tracking		
Journals		
Institutional changes/implementation		
Math majors: progress after calculus/tracking		
Graduation information		
Other _____		

VII. Summary of results: paragraph

* General success of efforts: none, low, medium, high
* Highlights of positive/negative attributes of project
* Other _____

VIII. Perceived quality of evaluation (circle one based on number of indicators that apply; number needed for each rating is in parentheses)

none (0)   low (1-2)   medium (3-4)   high (5+)

Indicators of evaluation quality (circle all that apply)

instruments reviewed by peers
advisory committee
consistency in evaluation plan
defined goals prior to evaluation
multiple measures used
evaluation plan disseminated
appropriate use of statistics
other _____

tested measures for validity/reliability
external evaluator
developed and used evaluation plan
controlled for confounding variables
results published in refereed journal
quantitative *and* qualitative measures
multi-year evaluation

IX. Dissemination efforts (circle all that apply; include number where indicated)

		Completed?	
Planned activity	project (#)	evaluation (#)	total (#)

journal articles
invited colloquia
info. requests from other institutions
use at other institutions
conference presentations
organized workshops
publication of course materials
software development
textbooks
other _____

X. Other information/unexpected findings

Calculus Reform
Perceived Quality of Evaluation

Definitions/Examples for Indicators of Evaluation Quality

1. Instruments reviewed by peers
2. Advisory committee
3. Consistency in evaluation plan: Example of not being consistent—no pre-testing although there was post-testing, i.e., lacks baseline data
4. Defined goals/outcomes (in the proposal) prior to evaluation
5. Multiple measures used: Example—faculty *and* student surveys used
6. Evaluation plan disseminated
7. Appropriate use of statistics: Example—use of statistics on course grades would not be appropriate. Furthermore, simply citing figures (e.g., 10 more student did better or 30% did better) would not be appropriate. When in doubt, give it to them and make a note in the Additional Information section.
8. Tested measures for validity/reliability: Used other people's stuff (e.g., well-known attitudinal surveys); Tested on sample/pilot group.
9. External evaluator
10. Developed and used evaluation plan—Did they follow through? Checks in both columns must match
11. Controlled for confounding variables
12. Evaluation results published in refereed journal
13. Quantitative *and* qualitative measures
14. Multi-year evaluation

August 16, 1996

Professor Jane Doe
Any University
Anytown, USA

Dear Professor Doe:

It is important that the academic community reflect upon the accomplishments and challenges of the efforts in calculus reform over the past decade. More importantly, we must determine the various outcomes and patterns of reform in order to develop an appropriate focus for future efforts. A project of the American Educational Research Association (AERA) Research Fellowship Program, supported by NSF, will compile and document studies on the impact and nature of calculus reform. To initiate this process, your cooperation in collecting information would be appreciated.

Specifically, please send me available materials on the evaluation of calculus reform efforts in which you or your colleagues have been involved, including published or unpublished papers that document your efforts and data collection instruments that have been developed as part of these studies. Studies may be quantitative or qualitative in nature and may include such information as student performance in calculus and subsequent courses, attrition rates, student and faculty attitudes, data from interviews and surveys, changes in departmental and institutional cultures, financial support generated within your institution and from industry, effects on curriculum and course development, new modes of assessing student learning, and other important indicators. Descriptive information that discusses the nature of your calculus reform efforts should also be included. In addition to evaluation documentation, we would also appreciate the names and addresses of other individuals you recommend we contact. These contacts may certainly include individuals involved in the evaluation of projects not funded by NSF.

The evaluation information you provide will be synthesized for a report to be distributed in summer 1997. The report will not identify any particular project, but will focus on the conclusions that can be made about the calculus reform movement as discerned from the documents submitted. The results of this project will contribute to the NSF program committed to evaluating educational innovations and to other efforts to monitor the developments in calculus, such as a recent project of Mathematicians and Education Reform (MER). This report is also expected to promote continued discussions about evaluating the efforts in calculus. NSF's Evaluation Program is devoted to supporting evaluation efforts and you are encouraged to consider this program when developing such projects. For further information, you may contact Conrad Katzenmeyer, Senior Program Director for Evaluation, at (703) 306-1655 X5812 or via email at ckatzenm@nsf.gov or visit the NSF Web site at http://red.www.nsf.gov.

The calculus reform movement has had a significant national impact on the teaching and learning of calculus, as well as mathematics in general. Your continued efforts and leadership in understanding the nature and outcomes of these efforts, as well as supporting continued reform of mathematics education, are very much needed. Your contributions and suggestions for this project should be sent by October 31, 1996 to

Susan L. Ganter
AERA Fellow
National Science Foundation
Suite 855
4201 Wilson Blvd.
Arlington, VA 22230.

We hope that this evaluation project will contribute to future efforts in calculus and appreciate the information you provide. A copy of the report will be sent to you upon its completion. Please contact me if you have any questions or comments at (703) 306-1655 X5813 or by email at sganter@nsf.gov.

Sincerely,

Susan L. Ganter
Research Fellow of the American Educational Research Association
Division of Research, Evaluation and Communication

cc: James H. Lightbourne, III
    Division of Undergraduate Education

Calculus Evaluation Database
National Objectives of Reform
Definitions

cooperative learning—any activities in which students work together in class or on assignments; other terms to look for: group learning, collaborative learning

technical writing—any writing requirements that are part of the reform course; other terms to look for: writing, student reports

graphing calculators—use of graphing calculators as an integral part of the course (simply allowing their use does not count here); other terms to look for: TI-82, etc.

problem solving skills—activities that focus on students' ability to formulate and solve mathematical problems; other terms to look for: critical thinking

conceptual understanding—activities that help students to understand and solve problems beyond rote manipulation; emphasis on concepts of calculus, rather than algorithms; other terms to look for: ability to discuss calculus, non-traditional answers to problems

open-ended problems—problems that have more than one correct solution; other terms to look for: multiple solutions

applications—problems that utilize information from other disciplines; other terms to look for: real-world problems; modeling

real-world modeling—ability to formulate a problem in mathematical language; often related to applications; other terms to look for: applications, real-world problems, setting up solutions

alternative assessment—*any* method used to assess student progress as part of the course grade that is different from traditional testing; other terms to look for: lab assignments, projects, journals, writing

discovery learning—activities in which students work individually or in groups to learn about calculus concepts; other terms to look for: Socratic method, group work, non-lecture format

extended-time projects—assigned problems that are completed over a time period of several days to several weeks; usually completed in teams; other terms to look for: group projects, extended lab assignments, report writing, group presentations

laboratory experience—activity that is part of class time in which the students are in a laboratory setting; this is usually a computer lab, but does not have to be; other terms to look for: computer lab

differential equations—the incorporation of differential equations as part of the calculus course; usually for the purpose of introducing and solving applications; other terms to look for: ODE, DE

oral presentations—in-class presentations made by students; other terms to look for: student presentations

computer use—any required use of computers in or out of class; this also includes use of computer in classroom demonstrations; <u>other terms to look for</u>: technical demos, technology, CAS, Maple, Mathematica, Derive, Theorist

mathematical reading—any reading that is required as part of the course, including specific reading from the textbook or other sources; <u>other terms to look for</u>: reading, student research

approximation—the use of problems that encourage students to approximate and estimate as part of the problem-solving process; <u>other terms to look for</u>: estimation, numerical solutions

multiple representations—the presentation of problems through different representations, specifi-cally, symbolic, verbal, numerical, and graphical; <u>other terms to look for</u>: graphs, data, tables, statistics, charts

References

Ahmadi, D. C. (1995). A Comparison Study between a Traditional and Reform Calculus II Program. Unpublished doctoral dissertation, The University of Oklahoma.

Alexander, E. H. (1997). An Investigation of the Results of a Change in Calculus Instruction at the University of Arizona. Unpublished doctoral dissertation.

Allen, G. H. (1995). A Comparison of the Effectiveness of the Harvard Calculus Series with the Traditionally Taught Calculus Series. Unpublished master's thesis, San Jose State University.

American Mathematical Association of Two Year Colleges (1995). *Crossroads in Mathematics: Standards for introductory college mathematics before calculus.* Memphis, TN: Author.

Andrews, G. E. (1997). Personal communication with the author.

Andrews, G. E. (1996). Mathematical Education: A case for balance. *The College Mathematics Journal, 27(5)*, 341–48.

Andrews, G. E. (1995). The Irrelevance of Calculus Reform: Ruminations of the sage-on-the-stage. *UME Trends, 6(6),* 17 & 23.

Armstrong, G., Garner, L., & Wynn, J. (1994). Our Experience with Two Reformed Calculus Programs. *Problems, Resources, and Issues in Mathematics Undergraduate Studies, 4(4)*, 301–11.

Asiala, M., Brown, A., DeVries, D., Dubinsky, E., Mathews, D., & Thomas, K. (1996). A Framework for Research and Curriculum Development in Undergraduate Mathematics Education. In E. Dubinsky, A. H. Schoenfeld, & J. Kaput (Eds.), *Research in Collegiate Mathematics Education, 2* (pp. 1–32). Providence, RI: American Mathematical Society.

Aspinwall, L. N. (1994). The Role of Graphic Representation and Students' Images in Understanding the Derivative in Calculus: Critical case studies. Unpublished doctoral dissertation, The Florida State University.

Association for Women in Mathematics (1989). Gender Differences in Mathematical Ability—Perception vs. Performance. *AWM Newsletter, 19(4).* College Park: Author.

Atkins, C. D., Jr. (1994). A Study to Produce Guidelines for Evaluating Calculus Reform Projects. Unpublished doctoral dissertation, North Carolina State University at Raleigh.

Barton, S. D. (1995). Graphing Calculators in College Calculus: An examination of teachers' conceptions and instructional practice. Unpublished doctoral dissertation, Oregon State University.

Basil, G. J. (1974). The Effects of Writing Computer Programs on Achievement and Attitude in Elementary Calculus. *Dissertation Abstracts International, 35(4)*, 2114A–2115A.

Baum, J. D. (1958). Mathematics, Self-taught. *American Mathematical Monthly, 65(9),* Washington, DC: Mathematical Association of America, 701–05.

Bell, F. H. (1970). A Study of the Effectiveness of a Computer-oriented Approach to Calculus. *Dissertation Abstracts International, 31,* 1096A.

Bonsangue, M. (1994). *An Efficacy Study of the Calculus Workshop Model.* Unpublished Manuscript, California State University, Fullerton.

Bookman, J. (2000). Program Evaluation and Undergraduate Mathematics Renewal: The impact of calculus reform on student performance in subsequent courses. In S. L. Ganter (Ed.), *Calculus Renewal: Issues for undergraduate mathematics education in the next decade* (pp. 91–102). New York: Kluwer Academic/Plenum Publishers.

Bookman, J. & Friedman, C. P. (1998). Student Attitudes and Calculus Reform. *School Science and Mathematics* (March), 117–122.

Bookman, J. & Friedman, C. P. (1994). A Comparison of the Problem Solving Performance of Students in Lab Based and Traditional Calculus. In E. Dubinsky, A. H. Schoenfeld, & J. Kaput (Eds.), *Research in Collegiate Mathematics Education, 1* (pp. 101–16). Providence, RI: American Mathematical Society.

Boyer, E. L. (1990). *Scholarship reconsidered: Priorities for the professoriate.* Princeton, NJ: Carnegie Foundation for the Advancement of Teaching.

Boyer, E. L. (1987). *College: The undergraduate experience in America.* New York, NY: Harper & Row.

Brechting, M. C. & Hirsch, C. (1977). The Effects of Small Group-Discovery Learning on Student Achievement and Attitudes in Calculus. *AMATYC, 11,(2),* pp. 77–82.

Brunett, M. R. (1996). Assessing Calculus Reform at a Two-year College. Evaluation report prepared for Montgomery College, Rockville, MD. Summary of "A Comparison of Problem Solving Abilities Between Reform Calculus Students and Traditional Calculus Students." *Dissertation Abstracts International, 57,* 1A.

Buccino, A. (2000). Politics and Professional Beliefs in Evaluation: The case of calculus renewal. In S. L. Ganter (Ed.), *Calculus Renewal: Issues for undergraduate mathematics education in the next decade* (pp. 121–146). New York: Kluwer Academic/Plenum Publishers.

Buck, R. C. (1962). Teaching Machines and Mathematics Programs: Statements by R. C. Buck. *American Mathematical Monthly, 69(6),* Washington, DC: Mathematical Association of America, 561–64.

Burgess (1996). Report on Implementation of Harvard Calculus Curriculum at Paradise Valley Community College. *Focus on Calculus.* New York, NY: John Wiley & Sons, Inc.

Carruthers, C. (1996). *Experience with Reform Calculus 1994–1996.* Evaluation report for Scottsdale Community College submitted to the National Science Foundation.

Chronicle of Higher Education (1997). The Effects of a Decade of Reform on how Calculus is Taught. *Chronicle of Higher Education, 43(7),* letters to the editor, B3, 10.

Cipra, B. (1996). Calculus Reform Sparks a Backlash. *Science, 271,* Washington, DC: American Association for the Advancement of Science, 901–902.

Cipra, B. A. (1988). Recent Innovations in Calculus Instruction. In L. A. Steen (Ed.) *Calculus for a New Century.* Mathematical Association of America: Washington, DC pp. 95–103.

Collins, R. J. (1996). *Calculus Revitalization at Xavier University.* Summary of paper presented at the Ohio Sectional Meeting of the Mathematical Association of America, March, 1992, Dayton, OH.

Committee on the Undergraduate Program in Mathematics (1969). Calculus with Computers. *Newsletter of the CUPM, 4* (August). Washington, DC: Mathematical Association of America.

Confrey, J. (1980). Conceptual Change, Number Concepts, and the Introduction to Calculus. Unpublished doctoral dissertation, Cornell University, Ithaca, NY.

Cope, C. L. (1980). A Comparison of a Large Lecture Format Featuring Small Group Discussions with a Traditional Lecture Test Format in a Decision Mathematics Course. *Dissertation Abstracts International, 41(3),* 973A.

Cribbs, et al. (1996). Report on Implementation of Harvard Calculus Curriculum at City College of San Francisco. *Focus on Calculus*. New York, NY: John Wiley & Sons, Inc.

Crocker, D. A. (1991). A Qualitative Study of Interactions, Concept Development and Problem-Solving in a Calculus Class Immersed in the Computer. Doctoral Dissertation, The Ohio State University. *Dissertation Abstracts International, 52A*, pp. 2850–2851.

Crockett, C. & Kiele, W. (1993). A Comprehensive Game Plan for Calculus Reform: Thoughts grown through experience. *Problems, Resources, and Issues in Mathematics Undergraduate Studies, 3(4)*, 355–70.

Cronbach, L. (1963). Course Improvement Through Evaluation. *Teachers College Record, 64*, 672–83.

Culotta, E. (1992). The Calculus of Education Reform. *Science, 255*, 1060–62.

Cummins, I. (1960). A Student Experience-Discovery Approach to the Teaching of Calculus. *The Mathematics Teacher, 53(3)*, Reston, VA: National Council of Teachers of Mathematics, 162–70.

Davidson, N. A. (Ed.,1990). Cooperative Learning in Mathematics: A handbook for teachers. Menlo Park, CA: Addison-Wesley Publishing Company.

Davidson, N. A. (1985). Small Group Learning and Teaching in Mathematics: A selective review of the research. In R. Slavin (Ed.) *Learning to Cooperate, Cooperating to Learn* (pp. 211–30). New York: Plenum Press.

Davidson, N. A. (1970). The Small Group-discovery Method of Mathematics Instruction as Applied in Calculus. Unpublished doctoral dissertation, The University of Wisconsin, Madison.

Davis, C. S., et al. (1996). *The Equity Equation: Fostering the advancement of women in the sciences, mathematics, and engineering*. San Francisco: Jossey-Bass Publishers.

Davis, P. W. (2000). Calculus Renewal and the World of Work. In S. L. Ganter (Ed.), *Calculus Renewal: Issues for undergraduate mathematics education in the next decade* (pp. 41-52). New York: Kluwer Academic/Plenum Publishers.

Davis, P. W. (1994). Mathematics in Industry: The job market of the future. *1994 SIAM Forum Final Report*. Philadelphia: Society for Industrial and Applied Mathematics.

Davis, P. W. (1993). Some Glimpses of Mathematics in Industry. *Notices of the American Mathematical Society, 40(7)*, Providence, RI: American Mathematical Society, 800–02.

Davis, T. A. (1966). An Experiment in Teaching Mathematics at the College Level by Programmed Instruction. *American Mathematical Monthly, 73(6)*, Washington, DC: Mathematical Association of America, 656–59.

DeBoar, D. D. (1974). A Comparison Study of the Effects of a Computer-oriented Approach to Calculus. *Dissertation Abstracts International, 34*, 3912B.

Donaldson, J. A. (1993). Report on Calculus I, 1993 Spring Semester. *Reports From Some Implementers*. Included in C[4]L Project information package. Unpublished.

Douglas, R. (1987). *Toward a Lean and Lively Calculus*. Report of the Tulane Conference on Calculus. Washington D.C.: Mathematical Association of America.

Dossey, J. (Ed., 1998). *Confronting the Core Curriculum: Considering change in the undergraduate mathematics major*, MAA Notes #45, Washington, DC: Mathematical Association of America.

Dreyfus, T. & Eisenberg, T. (1984). Intuitions on Functions. *Journal of Experimental Education, 52(2)*, pp. 77–85.

Dreyfus, T. & Eisenberg, T. (1982). Intuitive Functional Concepts: A baseline study on intuitions. *Journal for Research in Mathematics Education, 13(5)*, pp. 360–389.

Dubinsky, E. (1999). Assessment in One Learning Theory Based Approach to Teaching: A discussion. In B. Gold, S. Z. Keith, & W. A. Marion (Eds.) *Assessment Practices in Undergraduate Mathematics*, MAA Notes #49 (pp. 233–36), Washington, DC: Mathematical Association of America.

Dubinsky, E. (1992). A Learning Theory Approach to Calculus. In Z. Karian (Ed.) *Symbolic Computation in Undergraduate Mathematics Education*, MAA Notes #24 (pp. 48–55), Washington, DC: Mathematical Association of America.

Dudley, U. (1997). Is Mathematics Necessary? *The College Mathematics Journal, 25(5),* Washington, DC: Mathematical Association of America, 364.

Durell, F. (1894). Application of the New Education to the Differential and Integral Calculus. *American Mathematical Monthly, 1(1) & 1(2),* Washington, DC: Mathematical Association of America, 15–19, 37–41.

Dyer-Bennett, J., Fuller, W. R., Seibert, W. E., & Shanks, M. E. (1958). Teaching Calculus by Closed-circuit Television. *American Mathematical Monthly, 65(6),* Washington, DC: Mathematical Association of America, 430–39.

Eiseman, J. W., Fairweather, J. S., Rosenblum, S., & Britton, E. (1996). *Evaluation of the Division of Undergraduate Education's Course & Curriculum Development Program: Case study summaries.* Prepared under contract for the National Science Foundation, Andover, MA: The Network, Inc.

Eisner, E. (1975). *The Perceptive Eye: Toward the reformation of educational evaluation.* Presentation to the Stanford Evaluation Consortium. Stanford: December, 1975.

Eisner, E. (1969). Instructional and Expressive Objectives: Their formulation and use in curriculum. *AERA Monograph Series in Curriculum Evaluation, No. 3.* Chicago: Rand McNally.

Ellis, W., Jr. (2000). Technology and Calculus. In S. L. Ganter (Ed.), *Calculus Renewal: Issues for undergraduate mathematics education in the next decade* (pp. 53-68). New York: Kluwer Academic/Plenum Publishers.

Ferrini-Mundy, J. & Graham, K. G. (1991). Research in Calculus Learning: Understanding of Limits, Derivatives, and Integrals. In J. J. Kaput and E. Dubinsky (Eds.) *Research Issues in Undergraduate Mathematics Learning: Preliminary Analyses and Results.* MAA Notes, #33. Mathematical Association of America: Washington, DC.

Ferrini-Mundy, J. & Lauten, D. (1994). Learning about Calculus Learning. *The Mathematics Teacher, 87(2),* Reston, VA: National Council of Teachers of Mathematics, 115–21.

Fiedler, L. A. (1969). A Comparison of Achievement Resulting from Learning Mathematical Concepts by Computer Programming Versus Class Assignment Approach. *Dissertation Abstracts International, 29,* 3910A.

Flores, A. (1985). Effect of Computer Programming on the Learning of Calculus Concepts. *Dissertation Abstracts International, 46(12),* 3640A.

Foley, G. & Ruch, D. (1995). Calculus Reform at Comprehensive Universities and Two-Year Colleges. *UME Trends, 6(6),* 8–9.

Francis, E. J. (1992). The Concept of Limit in College Calculus: Assessing student understanding and teacher beliefs. Unpublished doctoral dissertation, University of Maryland, College Park.

Frid, S. (1994). Three Approaches to Undergraduate Instruction: Their nature and potential impact on students' language use and sources of conviction. In E. Dubinsky, A. H. Schoenfeld, & J. Kaput (Eds.), *Research in Collegiate Mathematics Education, 1* (pp. 69–100), Providence, RI: American Mathematical Society.

Ganter, S. L. (Ed., 2000). Calculus Renewal: Issues for undergraduate mathematics education in the next decade. New York: Kluwer Academic/Plenum Publishers.

Ganter, S. L. (1999). An Evaluation of Calculus Reform: A preliminary report of a national study. In B. Gold, S. Z. Keith, and W. A. Marion (Eds.), *Assessment Practices in Undergraduate Mathematics,* MAA Notes #49 (pp. 233–36), Washington, DC: Mathematical Association of America.

Ganter, S. L. (1997a). A Report on Evaluation in Calculus Reform. *Mathematics Education Research Newsletter*. Chicago: Mathematics and Education Reform.

Ganter, S. L. (1997b). Impact of Calculus Reform on Student Learning and Attitudes, *AWIS Magazine, 26(6)*, Washington, DC: Association for Women in Science, 10–15.

Ganter, S. L. (1994). The Importance of Empirical Evaluations of Mathematics Programs: A case from the calculus reform movement. *Focus on Learning Problems in Mathematics, 16(2)*. Framingham, MA: Center for Teaching/Learning of Mathematics.

Ganter, S. L., & Jiroutek, M. R. (2000). The Need for Evaluation in the Calculus Reform Movement: A comparison of two calculus teaching methods. In E. Dubinsky, A. Schoenfeld, and J. Kaput (Eds.), *Research in Collegiate Mathematics Education, IV* (42–62), Providence, RI : American Mathematical Society.

George, M. D. (2000). Calculus Renewal in the Context of Undergraduate SMET Education. In S. L. Ganter (Ed.), *Calculus Renewal: Issues for undergraduate mathematics education in the next decade* (pp. 1-10). New York: Kluwer Academic/Plenum Publishers.

Gordon, S. P. (2000). Renewing the Precursor Courses: New challenges, opportunities, and connections. In S. L. Ganter (Ed.), *Calculus Renewal: Issues for undergraduate mathematics education in the next decade* (pp. 69–90). New York: Kluwer Academic/Plenum Publishers.

Graham, K. G. & Ferrini-Mundy, J. (1989). *An Exploration of Student Understanding of Central Concepts in Calculus*. Paper presented at the annual meeting of the American Educational Research Association: San Francisco, CA.

Hamm, D. M. (1990). The Association Between Computer-Oriented and Non-Computer-Oriented Mathematics Instruction, Student Achievement, and Attitude Toward Mathematics in Introductory Calculus. Doctoral Dissertation, University of North Texas. *Dissertation Abstracts International, 50,* 2817A.

Hart, D. K. (1992). Building Concept Images: Supercalculators and students' use of multiple representations in calculus. Doctoral Dissertation, Oregon State University. *Dissertation Abstracts International, 52(12),* 4254A.

Hastings, N. B. (1997). The Workshop Mathematics Program: Abandoning lectures. In D'Avanzo, C & McNeal, A. (Eds.) *Student-active science: models of innovation in college science teaching*. Orlando, FL: Saunders College Publishing.

Haver, W. E. (1998). *Calculus: Catalyzing a national community for reform, NSF awards 1987–1995*. Washington, DC: Mathematical Association of America.

Hawker, C. M. (1986). The Effects of Replacing Some Manual Skills with Computer Algebra Manipulations on Student Performance in Business Calculus. Doctoral Dissertation, Illinois State University. *Dissertation Abstracts International, 47,* 2934A.

Heid, M. K. (1988). Resequencing Skills and Concepts in Applied Calculus Using the Computer as a Tool. *Journal for Research in Mathematics Education, 19(4),* Reston, VA: National Council of Teachers of Mathematics, 3–25.

Heid, M. K. (1984). An Exploratory Study to Examine the Effects of Resequencing Skills and Concepts in an Applied Calculus Curriculum Through the Use of the Microcomputer. Doctoral Dissertation, University of Maryland. *Dissertation Abstracts International, 46, 1548A.*

Herschberger, L. D. & Plantholt, M. (1994). Assessing the Harvard Consortium Calculus at Illinois State University. *Focus on Calculus: A newsletter for the calculus consortium based at Harvard University, (6),* 5.

Hiebert, J. & Carpenter, T. (1992). Learning and Teaching with Understanding. In D. Grouws (Ed.) *Handbook of Research on Mathematics Teaching and Learning*. Macmillan: New York, pp. 65–97.

Holden, L. S. (1967). Motivation for Certain Theorems of the Calculus. Unpublished doctoral dissertation, Ohio State University.

Holoein, M. O. (1971). Calculus and Computing: A comparative study of the effectiveness of computer programming as an aid in learning selected concepts in first-year calculus. *Dissertation Abstracts International, 31,* 4490A.

Hurley, E. A. (1983). Peer Teaching in a Calculus Classroom: The influence of ability. *Dissertation Abstracts International, 44(4),* 694A.

Hurley, J. F., Koehn, U., & Ganter, S. L. (1999). Effects of Calculus Reform: Local and national. *American Mathematical Monthly, 106(9),* Washington, DC: Mathematical Association of America, 800–11.

Jackson, H. E. (1979). A Comparison of the Effectiveness of Three Instructional Formats in Introductory Calculus on Student Achievement and Attrition. *Dissertation Abstracts International, 39(11),* 6606A.

Jackson, M. B. (1996). Personal correspondence with author about evaluation of calculus reform at Earlham College.

Johnson, C. (1997). Calculus Reform: An opinion. *School Science and Mathematics, 97(2),* letter to the editor.

Johnson, R., Johnson, R., & Stanne, M. (1986). Comparison of Computer-assisted, Cooperative, Competitive, and Individualistic Learning. *American Educational Research Journal, 23(3),* Washington, DC: American Educational Research Association, 382–92.

Joint Policy Board for Mathematics (1994). *Recognition and Rewards in the Mathematical Sciences.* Washington, DC: Author.

Judson, P. T. (1988). Effects of Modified Sequencing of Skills and Applications in Introductory Calculus. Doctoral Dissertation, University of Texas at Austin. *Dissertation Abstracts International, 49,* 1397A.

Keith, S. Z. (1995). How Do Students Feel about Calculus Reform, and How Can We Tell? *UME Trends, 6(6),* 6 & 31.

Keynes, H. B. (1991). The Calculus Curriculum Reform Movement: Some views from the outside. *Problems, Resources, and Issues in Mathematics Undergraduate Studies, 1(4),* 359–67.

Keynes, H. B., Olson, A, M., O'Loughlin, D. J., & Shaw, D. (2000). Redesigning the Calculus Sequence at a Research University: Faculty, professional development, and institutional issues. In S. L. Ganter (Ed.), *Calculus Renewal: Issues for undergraduate mathematics education in the next decade* (pp. 103–120). New York: Kluwer Academic/Plenum Publishers.

Klein, D. & Rosen, J. (1997). Calculus Reform—For the $millions. *Notices of the AMS, 44(10),* Providence, RI: American Mathematical Society, 1324–25.

Klopfenstein, K. F. (1977). The Personalized System of Instruction in Introductory Calculus. *American Mathematical Monthly, 84(2),* 120–24.

Kolata, G. B. (1987). Calculus Reform: Is it needed? Is it possible? In L. A. Steen (Ed.) *Calculus for a New Century.* Mathematical Association of America: Washington, DC, pp. 89–94.

Kroll, D. L. (1989). Cooperative Mathematical Problem Solving and Metacognition: A case study of three pairs of women. Unpublished doctoral dissertation, Indiana University.

Larsen, C. M. (1961). The Heuristic Standpoint in the Teaching of Elementary Calculus. Unpublished doctoral dissertation, Stanford University.

Lepowsky, W. L. (1996). *Instituting Calculus Reform: A community college-state university consortium model.* Final report to the National Science Foundation, NSF #9450735.

Levi, H. (1963). An Experimental Course in Analysis for College Freshmen. *American Mathematical Monthly, 70(8),* Washington, DC: Mathematical Association of America, 877–79.

Levine, J. L. (1968). A Comparative Study of Two Methods of Teaching Mathematical Analysis at the College Level. Unpublished doctoral dissertation, Columbia University.

Lightbourne, J. H., III (2000). Crossing the Discipline Boundaries to Improve Undergraduate Mathematics Education. In S. L. Ganter (Ed.), *Calculus Renewal: Issues for undergraduate mathematics education in the next decade*(pp. 147–158). New York: Kluwer Academic/Plenum Publishers.

Loftsgaarden, D. O., Rung, D. C., & Watkins, A. E. (1997). *Statistical Abstract of Undergraduate Programs in the Mathematical Sciences in the United States: Fall 1995 CBMS survey*. Washington, DC: Mathematical Association of America.

Loomer, N. J. (1976). A Multidimensional Exploratory Investigation in Small Group-heuristic and Expository Learning in Calculus. *Dissertation Abstracts International, 21(9)*, 2632–2633.

Lovelace, T. L. & McKnight, C. K. (1980). The Effects of Reading Instruction on Calculus Students' Problem Solving. *Journal of Reading, 23(4)*, 305–08.

Mathews, D. M. (1995). Time to Study: The C^{4}L experience. *UME Trends.*

May, K. O. (1964). *Programmed Learning and Mathematical Education.* San Francisco: The Committee on Educational Media of the Mathematical Association of America.

May, K. O. (1962). Small Versus Large Classes. *American Mathematical Monthly, 69(5),* Washington, DC: Mathematical Association of America, 433, 434.

McCallum, W. (2000). The Goals of the Calculus Course. In S. L. Ganter (Ed.), *Calculus Renewal: Issues for undergraduate mathematics education in the next decade* (pp. 11–22). New York: Kluwer Academic/Plenum Publishers.

McKeachie, W. J. (1954). Student-centered Versus Instructor-centered Instruction. *Journal of Educational Psychology, 45(3),*143–50.

McKeen, R. L. (1970). A Model for Curriculum Construction through Observations of Students Solving Problems in Small Instructional Groups. Unpublished doctoral dissertation, University of Maryland, College Park.

McLellan, J. A. & Dewey, J. (1895). *The Psychology of Number and its Applications to Methods of Teaching Arithmetic.* New York: D. Appleton.

Moise, E. E. (1965). Activity and Motivation in Mathematics. *American Mathematical Monthly, 72(4),* Washington, DC: Mathematical Association of America, 407–12.

Monk, G. (1989). *A Framework for Describing Student Understanding of Functions.* Paper presented at the Annual meeting of the American Educational Research Association, San Francisco, CA.

Monroe, H. L. (1966). A Study of the Effects of Integrating Analytic Geometry and Calculus on the Achievement of Students in these Courses. Unpublished doctoral dissertation, University of Pittsburgh.

Monteferrante, S. (1993). Implementation of Calculus, Concepts, and Computers at Dowling College. *Collegiate Microcomputer, 11, (2).*

Moore, W. K. (1976). An Investigation of the Effectiveness of an Audio-taped Presentation of a Unit in College Calculus. *Dissertation Abstracts International, 36(10)*, 2891B.

Morstad, D. P., Jr. (1996). *Computers and Calculus.* Final report submitted to the National Science Foundation, NSF #9451556.

Mumford, D. (1997). Calculus Reform for the Millions. *Notices of the American Mathematical Society, 44(5),* Providence, RI: American Mathematical Society, 559–563.

Murphy, J. E. (1975). Some Specific Suggestions for the Use of a Portable, Algebraic-language Computer in the Analytical Geometry and Calculus Classroom. *Dissertation Abstracts International, 36(5)*, 2280B.

Narasimhan, C. C. (1993). Calculus Reform for the Non-science Client Disciplines. *Problems, Resources, and Issues in Mathematics Undergraduate Studies, 3(3)*, 254–62.

National Research Council (1996). *From Analysis to Action: Undergraduate education in science, mathematics, engineering, and technology*. Washington, DC: National Academy Press.

National Research Council (1983). *A Nation at Risk*. Washington, DC: National Academy Press.

National Science Board (1998). *Science & Engineering Indicators—1998*. Arlington, VA: National Science Foundation.

National Science Board (1986). *Undergraduate Science, Mathematics and Engineering Education: Role for the National Science Foundation and recommendations for action by other sectors to strengthen collegiate education and pursue excellence in the next generation of U.S. leadership in science and technology*. Report of the Task Committee on Undergraduate Science and Engineering Education, H. Neal, Chair. Washington DC: National Science Foundation.

National Science Foundation (1996). *Shaping the Future: New expectations for undergraduate education in science, mathematics, engineering, and technology*. Report of the Advisory Committee for Review of Undergraduate Education, M. George, Chair. Arlington, VA: Author.

National Science Foundation (1995). *Restructuring Engineering Education: A focus on change*. Report of the NSF Workshop on Innovations in Engineering, C. Meyers, Chair. Arlington, VA: Author.

National Science Foundation (1993). *Innovation and Change in the Chemistry Curriculum*. Report of the NSF Workshop on Innovations in Chemistry, S. Ege and O. Chapman, Co-Chairs. Washington, DC: Author.

National Science Foundation (1991). *Undergraduate Curriculum Development: Calculus*. Report of the Committee of Visitors, P. Treisman, Chair. Washington, DC: Author.

National Science Foundation (1987). *Undergraduate Curriculum Development in Mathematics: Calculus*. Program announcement, Division of Mathematical Sciences. Washington, DC: Author.

Newell, J. C. (1994). Student Experiences in a First Semester University Calculus Course: A study using ethnographic methods. Unpublished master's thesis, Simon Fraser University (Canada).

Norfolk, T. S. (1997). A Mathematician's Apology? It's time to stop. *Focus, 17(1)*, Washington, DC: Mathematical Association of America, 14–15.

Orton, A. (1983a). Students' Understanding of Differentiation. *Educational Studies in Mathematics, 14(3)*, pp. 235–250.

Orton, A. (1983b). Students' Understanding of Integration. *Educational Studies in Mathematics, 14(1)*, pp. 1–18.

OuYang & Hann (1996). Report on Implementation of Harvard Calculus Curriculum at California State University, Hayward. *Focus on Calculus*. New York, NY: John Wiley & Sons, Inc.

Packel, E. (1996). *An Integrated Laboratory Environment for Majors in Mathematics*. Evaluation report submitted to the National Science Foundation.

Padgett, E. E., III (1994). Calculus I with a Laboratory Component. Unpublished doctoral dissertation, Baylor University.

Palmiter, J. R. (1991). Effects of Computer Algebra Systems on Concept and Skills Acquisition in Calculus. *Journal for Research in Mathematics Education, 22(2)*, 151–56.

Palmiter, J. R. (1986). The Impact of a Computer Algebra System on College Calculus. Doctoral Dissertation, The Ohio State University. *Dissertation Abstracts International, 47*, 1640A.

Pascarella, E. (1978). Interactive Effects of Prior Mathematics Preparation and Level of Instructional Support in College Calculus. *American Educational Research Journal, 15(2)*, pp. 275–285.

Peluso, A. & Baranchik, J. (1977). Self-paced Mathematics Instruction: A statistical comparison with traditional teaching. *American Mathematical Monthly, 85(2)*, 124–29.

Pence, B. (1996). Investigating Change in the Calculus Curriculum at San Jose State University. In *Proceedings of the Seventh Annual Conference on Technology in Collegiate Mathematics* (pp. 362-66). Boston: Addison-Wesley Publishing Company, Inc.

Penn, H. L. (1994). Comparison of Test Scores in Calculus I at the Naval Academy. *Focus on Calculus: A newsletter for the calculus consortium based at Harvard University, (6)*, pp. 6–7.

Peterson, I. (1987). Calculus Reform: Catching the wave? *Science News, 132(19)*, p. 317.

Pinter-Lucke, C. (1996). *Evaluation of Calculus Reform at California State Polytechnic University, Pomona.* Report to the National Science Foundation.

Pólya, G. (1957). *How to Solve it: A new aspect of mathematical method.* (2nd Ed.). Princeton University Press: Princeton, NJ.

Porzio, D. T. (1994). The Effects of Differing Technological Approaches to Calculus on Students' Use and Understanding of Multiple Representations When Solving Problems. Unpublished doctoral dissertation, The Ohio State University.

Prettyman, T. (1996). *Classroom and Student Demonstrations for Active Learning in Introductory Mathematica and Computer Science Courses: Year one.* Evaluation report for Essex Community College submitted to the National Science Foundation.

Pustejovsky, S. (1996). Calculus Reform at Alverno College. Report to the National Science Foundation.

Ratay, G. M. (1994). Test Site Report: Success with lean and lively calculus. *Focus on Calculus: A newsletter for the calculus consortium based at Harvard University, (6)*, p. 7.

Ratay, G. M. (1993). Student Performance With Calculus Reform at the United States Merchant Marine Academy. *Primus, 3(1)*, pp. 107–111.

Ratay, G. M. (1992). Test Site Report: Success with lean and lively calculus. *Focus on Calculus: A newsletter for the calculus consortium based at Harvard University, (1)*, p. 4.

Reynolds, B. E. (1991). Calculus Reform: Implementation of a new curriculum. *Problems, resources, and Issues in Mathematics Undergraduate Studies, 1(4)*, 373–78.

Roberts, W. (Ed., 1996). *Calculus: The dynamics of change.* Washington, DC: Mathematical Association of America.

Rochowicz, J. A., Jr. (1993). An Analysis of the Perceived Impact of Computing Devices on Calculus Instruction in Engineering Curricula. Unpublished doctoral dissertation, Lehigh University.

Salem, A., Shorter, P., & Koelzer, J. (1996). *Rockhurst College Project CALC Implementation.* Evaluation report to the National Science Foundation.

Schneider, E. M. (1995). Testing the Rule of Three: A formative evaluation of the Harvard based calculus consortium curriculum. Unpublished doctoral dissertation, The University of Texas at Austin.

Schoenfeld, A. H. (1996). *Student Assessment in Calculus.* A Report of the NSF Working Group on Assessment in Calculus. Arlington, VA: National Science Foundation.

Scriven, M. (1991). *Evaluation Thesaurus, Fourth Edition.* Newbury Park: Sage Publications.

Scriven, M. (1967). The Methodology of Evaluation. *AERA Monograph Series in Curriculum Evaluation, No. 1.* Chicago: Rand McNally.

Schrock, C. S. (1989). Calculus and Computing: An exploratory study to examine the effectiveness of using a computer algebra system to develop increased conceptual understanding in a first-semester calculus course. Doctoral Dissertation, Kansas State University. *Dissertation Abstracts International, 50(7),*1926A.

Selden, J., Selden, A. & Mason, A. (1989). Can Average Calculus Students Solve Nonroutine Problems? *Journal of Mathematical Behavior, 8(2)*, pp. 45–50.

Shah (1996). Report on Implementation of Harvard Calculus Curriculum. *Focus on Calculus*. New York, NY: John Wiley & Sons, Inc.

Shelton, R. M. (1965). A Comparison of Achievement Resulting from Teaching the Limit Concept in Calculus by Two Different Methods. Unpublished doctoral dissertation, University of Illinois.

Silverberg, J. (1994). *A Longitudinal Study of an Experimental vs. Traditional Approach to Teaching First Semester Calculus*. Unpublished manuscript.

Simonsen, L. M. (1995). Teachers' Perceptions of the Concept of Limit, the Role of Limits, and the Teaching of Limits in Advanced Placement Calculus. Unpublished doctoral dissertation, Oregon State University.

Slavin, R. E. (1999). A Rejoinder: Yes, control groups are essential in program evaluation: A response to Pogrow. *Educational Researcher, 28(3)*, 36–38.

Slavin, R. E. (1985). An Introduction to Cooperative Learning Research. In R. Slavin (Ed.), *Learning to Cooperate, Cooperating to Learn* (pp. 5–15). New York: Plenum Press.

Slavin, R. E. (1983). When Does Cooperative Learning Increase Student Achievement? *Psychological Bulletin, 94*, 429–45.

Smith, D. A. (2000). Renewal in Collegiate Mathematics Education: Learning from research. In S. L. Ganter (Ed.), *Calculus Renewal: Issues for undergraduate mathematics education in the next decade* (pp. 23–40). New York: Kluwer Academic/Plenum Publishers.

Smith, D. A. (1996). Trends in Calculus Reform. In A. Solow (Ed.), *Preparing for a New Calculus*, MAA Notes #36 (pp. 3–13). Washington, DC: Mathematical Association of America.

Smith, D. A. (1970). A Calculus-With-Computer Experiment. *Educational Studies in Mathematics, 3(1)*, 1–11.

Solow, A. (1996). *Preparing for a New Calculus*, MAA Notes #36. Washington, DC: Mathematical Association of America.

Soto-Johnson, H. (1996). Technological vs. Traditional Approach in Conceptual Understanding of Series. Unpublished manuscript.

Stannard, W. A. (1966). The Effect on Final Achievement in a Beginning Calculus Course Resulting from the Use of Programmed Materials Written to Supplement Regular Classroom Instruction. Unpublished doctoral dissertation, Montana State University.

Steen, L. A. (Ed., 1987). *Calculus for a New Century*. Washington, DC: Mathematical Association of America.

Stevens, F., Lawrenz, F., & Sharp, L. (1993). *User-Friendly Handbook for Project Evaluation: Science, mathematics, engineering and technology education*. Washington, DC: National Science Foundation.

Stockton, D. S. (1960). An Experiment with a Large Calculus Class. *American Mathematical Monthly, 67(10)*, Washington, DC: Mathematical Association of America, 1024–25.

Struik, R. R. & Flexer, R. J. (1977). Self-paced Calculus: A preliminary evaluation. *American Mathematical Monthly, 84(2)*, 129–34.

Tall, D. O. (1990). Inconsistencies in the Learning of Calculus and Analysis. *Focus on Learning Problems in Mathematics, 12(3 & 4)*, 49–63.

Tall D. O., et al. (1984). The Mathematics Curriculum and the Micro. *Mathematics in School, 13(4)*, 7–9.

Tall, D. O. (1978). The Dynamics of Understanding Mathematics. *Mathematics Teaching, 84*, 50–52.

Tall, D. O. (1977). Conflicts and Catastrophes in the Learning of Mathematics. *Mathematical Education for Teaching, 2(4)*, 2–18.

Tall, D. O. & Schwarzenberger, R. L. (1978). Conflicts in the Learning of Real Numbers and Limits. *Mathematics Teaching, 82*, pp. 44–49.

Tall, D. O. & Vinner, S. (1981). Concept Image and Concept Definition in Mathematics with Particular Reference to Limits and Continuity. *Educational Studies in Mathematics, 12(2)*, pp. 151–169.

Taylor, V. S. K. (1977). A Longitudinal Comparison of the Effect of an Individualized Calculus Course on the Cumulative Gradepoint Averages of First-year College Students. *Dissertation Abstracts International, 38(1)*, 143–144A.

Teles, E. (1992). Calculus Reform: What was happening before 1986. *Primus, 2(3)*, pp. 224–234.

Thompson, S. B. (1979). An experiment in individualized mastery learning of college calculus. *Dissertation Abstracts International, 39(8)*, 4795A.

Tidmore, E. (1994). A Comparison of Calculus Materials Used at Baylor University. *Focus on Calculus: A newsletter for the calculus consortium based at Harvard University, (6)*, 8.

Treisman, P. U. (1985). *A Study of the Mathematics Performance of Black Students at the University of California, Berkeley*. Unpublished doctoral dissertation, University of California, Berkeley.

Tucker, A. C. & Leitzel, J. R. C. (Eds., 1995). *Assessing Calculus Reform Efforts: A report to the community*. Washington, DC: Mathematical Association of America.

Tucker, T. W. (Ed.,1990). *Priming the Calculus Pump: Innovations and resources*. Washington, DC: Mathematical Association of America.

Tufte, F. W. (1990). The Influence of Computer Programming and Computer Graphics on the Formation of the Derivative and Integral Concepts (Derivative Concepts). *Dissertation Abstracts International, 51/04*, 1149A.

Turner, V. D., Alders, C. D., Hatfield, F., Croy, H., & Sigrist, C. (1966). A Study of Ways of Handling Large Classes in Freshman Mathematics. *American Mathematical Monthly, 73(7)*, Washington, DC: Mathematical Association of America, 768–70.

Urion, D. K. & Davidson, N. A. (1992). Student Achievement in Small-group Instruction versus Teacher-centered Instruction in Mathematics. *Problems, Resources, and Issues in Mathematics Undergraduate Studies, 2(3)*, 257–64.

Utter, F. W. (1996). Relationships Among AP Calculus Teachers' Pedagogical Content Beliefs, Classroom Practice, and Their Students' Achievement. Unpublished doctoral dissertation, Oregon State University.

Valverde, L. (1984). Underachievement and Underrepresentation of Hispanics in Mathematics and Mathematics-related Careers. *Journal for Research in Mathematics Education, 15*, Reston, VA: National Council of Teachers of Mathematics.

Waggoner, R. (1996). Personal correspondence with author about evaluation of calculus reform at University of Southwestern Louisiana.

Walker (1996). Report on Implementation of Harvard Calculus Curriculum at College of Alameda. *Focus on Calculus: A newsletter for the calculus consortium based at Harvard University*. New York, NY: John Wiley & Sons, Inc.

Wang, T. (1996). Personal correspondence with author about evaluation of calculus reform at Oakton Community College.

Webb, N., Ender, P., & Lewis, S. (1986). Problem Solving Strategies and Group Processes in Small Groups Learning Computer Programming. *American Educational Research Journal*, 23(2), Washington, DC: American Educational Research Association, 243-62.

Webber, R. P. (1996). Personal correspondence with author about evaluation of calculus reform at Longwood College.

Wells, P. J. (1995). Conceptual Understanding of Major Topics in First Semester Calculus: A study of three types of calculus courses at the University of Kentucky. Unpublished doctoral dissertation, University of Kentucky.

West, R. D. (1994). Evaluating the Effects of Changing a Mathematics Core Curriculum. Presentation at the Joint Mathematics Meeting (January), Cincinnati, OH.

White, R. M. (1987). Calculus of Reality. In L. A. Steen (Ed.) *Calculus for a New Century: A pump not a filter.* Mathematical Association of America: Washington, DC, pp. 6–9.

Williams, C. G. (1994). Using Concepts Maps to Determine Differences in the Concept Image of Function held by Students in Reform and Traditional Calculus. Unpublished doctoral dissertation, University of California, Santa Barbara.

Wilson, R. (1997). A Backlash Against 'Reform Calculus.' *Chronicle of Higher Education, 43(22)*, A12–A13.

Yackel, E., Cobb, P., & Wood, T. (1991). Small-group Interactions as a Source of Learning Opportunities in Second-grade Mathematics. *Journal for Research in Mathematics Education, 22(5),* Reston, VA: National Council of Teachers of Mathematics, 390–408.

Zandieh, M. (1997). The Evolution of Student Understanding of the Concept of Derivative. Unpublished doctoral dissertation, Oregon State University.

Zorn, P. (1991). Content and Discontent in San Antonio. *UME Trends, 3(1).*

DATE DUE

Demco, Inc. 38-293